Thank you for your
Great work in
Sustainability.
Thanks too for your
Passion for our
Planet!

Mark Edwards
Oct '09

# Green Algae Strategy

## End Oil Imports and
## Engineer Sustainable Food and Fuel

### Mark Edwards
GreenIndependence.org

**Key words:** algae, cyanobacteria, innovation, nanotechnology, biotechnology, sustainability, green solar energy, food, hunger, eflation, biofuels, pollution, ethanol, jet fuel, agriculture, medicines, pharmaceuticals, health, nutraceuticals, vaccines, reforestation, aquaculture, smoke death, environment, chemical engineering, business, social entrepreneur, social networks and collaboratory.

ISBN 1440421846

EAN-13 is 9781440421846

Tempe, Arizona

Independent Publisher Book Awards, IPPY.
http://www.independentpublisher.com/article.php?page=1231

**Copyright** © 2008, Mark R. Edwards, Rev. 7.2

*Green Algae Strategy* materials may be copied and used for educational purposes.

Cover photo – sushi and algal oil © 2008, Mark R. Edwards

# Dedication

To my wonderful life partner Ann Ewen and her passion for great food and loving support and to Sarah Edwards who finishes grace before family meals with "Please God, bless this food and help people who don't have food get some."

To Jacques Cousteau, who introduced and mentored my introduction to algae and global stewardship through his contributions to the U.S. Naval Academy where he helped initiate an oceanography program.

# Contents

Green Algae Strategy ................................................................. iii
Preface ...................................................................................... iii
Biowar I .................................................................................... vi
My Fabulous Friend from the Food Chain's Floor ................... iii

Chapter 1. What is Green Algae Strategy? ............................... 1
Chapter 2. What are the Global Challenges? ......................... 27
Chapter 3. Sustainable Food – Algaculture ............................ 57

Chapter 4. Why Algae? ........................................................... 77
Chapter 5. Algae Growth and Production .............................. 93
Chapter 6. Algal Production Challenges .............................. 113
Chapter 7. Products and Pollution Solutions ....................... 133

Chapter 8. Who is Producing Algae? .................................... 161
Chapter 9. Future Scenarios ................................................. 187
Chapter 10. What is Algae's Future? .................................... 211

Acknowledgements .............................................................. 228
Mark Edwards ...................................................................... 229
My Path to Green Independence ......................................... 230
Great Green Reading ........................................................... 230

# Green Algae Strategy Series by Mark Edwards

Biowar I, where food is burned for biofuel, must end with the withdrawal not of soldiers but ecologically damaging subsidies. The Green Algae Strategy series focuses on creating sustainable and affordable food and energy, SAFE production.

*Biowar I: Why Battles over Food and Fuel Lead to World Hunger.* 2007. The unintended consequences of producing corn ethanol on U.S. and world food markets will be catastrophic for U.S. fossil water, soils, air, food exports and food prices.

*Green Algae Strategy: End Oil Imports and Engineer Sustainable Food and Fuel.* 2008. Algae offer solutions for sustainable and affordable food and energy because algae are the most productive biomass source on Earth. Fossil agriculture is non-sustainable because it uses far too many non-renewable resources, especially cropland, fossil water, fossil fuels and diminishing agricultural chemicals.

*Green Solar Gardens: Algae's Promise to End Hunger.* 2008. Algaculture in small solar gardens distributed globally will enable SAFE production, locally. Solar gardens addresses the web of poverty and hunger including affordable food, fodder, fish food, fertilizer, fire for cooking and heating and fine medicines.

*Green Independence: End Oil Imports and Engineer Sustainable Food and Fuel.* 2009. (in press) Algae offer solutions for sustainable liquid transportation fuel because algae the most productive biomass source on Earth.

*Crash! The demise of Fossil Foods and the rise of Abundance.* 2009. Traditional fossil agriculture sits precariously on a foundation of unsustainable fossil resources the will become unaffordable and then run out. Abundant agriculture is sustainable because it uses plentiful inputs that are cheap and will not run out.

GAS content is available at http://GreenIndependence.org in color PDF for free download by students, teachers, scientists and food and energy policy leaders. The books are also available on Amazon.com.

# Preface

*Green Algae Strategy* shares the fascinating story of extraordinary innovation occurring not in deep space or in deep oceans but simply under our feet. Few people are aware that this simplest of organisms holds such great potential for desperately needed sustainable solutions for our very hungry, thirsty and needy planet.

The Green Algae Strategy engineers hope for a better life for billions of people who lack sufficient and affordable food, fresh water, fresh air, fertilizer and fuel for cooking and heating fires. The strategy includes cleaning polluted water and reforesting denuded land but those objectives are peripheral to the focus here on producing sustainable foods and biofuels. Since algae-based biofuels provide the strongest financial incentives for R&D, new food sources, pollution solutions, reforestation, medicines and other coproducts will all benefit from breakthroughs in algal production systems for biofuels.

Biotechnology applies science and engineering principles to living organisms to solve problems and to make useful products. Over the last century, many people and companies have lost fortunes trying to create commercial scale algal production. The laboratory studies are so promising, yet even modest scale field studies have typically become unmanageable, unstable and unproductive.

Advances in biotechnology, nanotechnology and chemical and mechanical engineering have changed the production landscape for algae from dismal to terrific. Green Algae Strategy lays out a roadmap for what may be the challenge of this century: solutions to sustainable food, water, pollution, reforestation and biofuels.

Algae will not be the silver bullet that singularly resolves sustainability issues. Truly renewable technologies that meet increasing world demand for food and energy will be solved by a portfolio approach that will include all renewable energy sources and biofuels. However, algae are poised to provide innovative, high value and engaging solutions.

Green Algae Strategy

## My Fabulous Friend from the Food Chain's Floor

**Imagine a humble single-celled organism,**
Too small to be visible to the human eye,
So old it was among the Earth's earliest life forms,
So delicate it may be carried on waves or wind,
So fragile it depends on sunshine, $CO_2$ and water for life.

**This organism**
Battled Earth's early fires, freezes and floods,
Endured droughts, deserts and destruction,
Survived volcanoes, glaciers and polar shifts,
Adapted to Earth's most extreme conditions.

**To survive over three billion years, this organism**
Learned how to hibernate through ecological crashes,
Discovered how to flourish when nourished,
Invented novel strategies to adapt to Mother Nature's whims,
Re-evolved independently in millions of harsh microclimates.

**This organism is responsible for**
Replacing Earth's early $CO_2$ atmosphere with $O_2$,
Enabling the evolution of land plants, insects and animals,
Serving as dinner for trillions; being the lowest rung on the food chain,
Providing over half our planet's oxygen today.

**It may be found:**
In all Earth's soils, waters, ice, altitudes and latitudes,
Developing in limitless shapes, sizes and structures,
Displaying a rainbows of colors, densities and shades,
Popping up in pools, ponds, wetlands, estuaries and oceans.

**This flexible organism may:**
Grow as one of more than 100,000 species,
Thrive as one of an infinite number of strains,
Develop incredibly fast or unbelievably slow,
Change its characteristics to adapt to nature or nurture.

**This organism provided:**
The first free lunch for bacteria, ameba, bugs, fish and frogs,
The green sheen of the well-known giant sloth's fur,
The fabulous flamingo pink that attracts flirting females,
The Royal Red dye for Cleopatra's gowns and Caesar's legions.

**It displays a passion for growth:**
A puddle from a desert thunderstorm may contain 100 types,
A rock under Antarctica's ice may exhibit 50 species,
Fossilized trees, plants and animals display hundreds of varieties,
And these long-dormant plant cells may still be viable after eons.

**This organism, compared with land plants:**
Grows many times faster because it grows 360°, not just 1°, up
Thrives in saltwater, brine, polluted or wastewater,
Blooms so fast it can triple its size in a single day,
Efficiently transforms sunshine into high energy biomass.

Green Algae Strategy

**This organism is very special:**
It needs no trunk – it floats in water,
It needs no stem – it grows in all directions,
It needs no leaves – it absorbs light directly,
It needs no growing season – it grows when the sun shines.

**This organism is an ecological master, it grows with:**
A zero carbon footprint – it converts $CO_2$ to $O_2$,
A low land footprint – it requires no cropland,
A low water footprint – its growing water can be recycled,
A small ecological footprint because it creates minimal waste.

**This organism:**
Feeds on greenhouse gasses from coal-fired plants,
Transforms industrial heat into plant oils and proteins,
Assimilates nitrogen pollution from agricultural waste,
Transforms industrial wastewater into clean water.

**This adaptive organism may grow to provide:**
A biomass of 60% proteins – energy used for foods,
Or a biomass with 60% lipids – oils used for biofuels,
Or with 90% carbohydrates – material to make paper
Very little waste, about 10%.

**This organism produces:**
Food 30 to 100 times faster than grains or grasses,
Biofuels many times faster and cheaper than corn,
Low cost medicines, pharmaceuticals and vaccines
Health foods, vitamins, minerals and fine cosmetics.

**This organism stumbles because people:**
See it gunk-up pools, ponds, aquariums and waterways,
Investigate, research and write mostly about how to kill it,
Are unaware pond stink comes from bacteria feeding on it,
Have mislabeled this plant as icky, yicky, smelly green slime.

**This organism is so humble it:**
Has prompted very little research and development,
Was dropped completely from U.S. government R&D for decades,
Has spawned only a handful of University labs in the U.S.,
Has realized far less than one percent of its potential.

**My friends, I want you to meet Green Magic –**
May I introduce to you to my fabulous friend
from the food chain's floor –
algae.

Mark Edwards, April 2008
Dedicated to Amos Richmond
For a rich life studying algae

**Algae**

 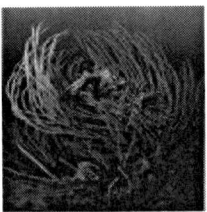 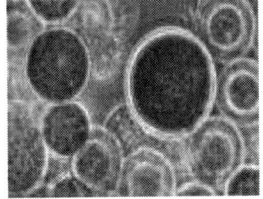

# Chapter 1. What is Green Algae Strategy?

Solutions for sustainable, sufficient and affordable global nutritious food and high-energy biofuel may lie not in deep space, deep oceans or deep underground pools but in simple mud puddles under our feet that contain probably 50 species of the fastest growing, energy rich biomass on Earth.

Green Algae Strategy engineers solutions to resolve a set of Earth's most intractable challenges including:

1. An end to oil imports
2. An end to American and global hunger
3. An end to the need to burn fossil fuels
4. Recapture of the carbon released in burned fossil fuels

Green Algae Strategy offers green independence for America and our global neighbors who will end the need oil imports and eventually for fossil fuels. Green independence takes advantage of nature's oldest, tiniest and yet fastest growing plant to recapture fossil carbon, to repair the Earth's atmosphere and to produce both food and biofuel.

Green solar, also known as algaculture and nanoculture, uses the sun's energy through photosynthesis to capture $CO_2$ near the Earth's surface, in water or air. Every pound of algae biomass sequesters 1.8 pounds of $CO_2$. Even though algae is the tiniest plant on Earth, representing only 0.5% of total plant biomass, algae create about 60% of the Earth's oxygen – more than all the forests and fields combined.

Green Algae Strategy

In the process of capturing carbon and producing pure oxygen, algae create harvestable green bioenergy and nutritious protein.

**Carbon neutral energy**

Green solar offers a sustainable, low cost and non-polluting source for creating carbon neutral biofuels in weeks instead of millions of years. Algae use abundant low cost inputs that are inexhaustible to capture carbon and create energy. Sunshine is free, waste water with nutrients are surplus from cities, businesses and farms and power and manufacturing plants have millions of tons of surplus $CO_2$.

Overconsumption of fossil fuels reduces fuel supplies and accelerates global warming from the release of greenhouse gases. Climate change jeopardizes food production, drives up the price of food and creates political instability. Climate change makes traditional agriculture non-sustainable for a variety of reasons, especially excess heat, drought, fierce storms and rising tides. Fortunately, algaculture has the potential to eliminate the need for fossil fuels by converting to truly renewable and sustainable food and fuels.

Global warming can be slowed by replacing fossil fuels with renewable carbon neutral alternatives. Green solar captures the sun's energy in plant biomass and converts it to socially useful energy and represents a clean alternative to fossil fuels. Green solar sequesters $CO_2$, improves ecosystems and does not disrupt traditional food production while producing both food and biofuel.

Traditional or yellow solar energy production employs panels with photovoltaic cells that absorb photons in sunlight with semiconductor materials such as silicon. Electrons are knocked loose from their atoms and flow through the semiconductor to produce electricity. Solar panels are relatively efficient at capturing about 33% of radiant solar energy and are carbon neutral. However, yellow solar does not convert energy to a liquid transportation fuel – only battery storage for transportation. Solar panels provide no coproducts, just electricity.

Green solar absorbs sunlight through photosynthesis and stores the energy in plant bonds that are the electron carriers. Algae synthesize water and $CO_2$ to convert sunshine to lipids that can be harvested as biodiesel. Algal communities grow in waste streams and thrive on wastewater, brine or saltwater. The non-lipid algal biomass, predominately protein and carbohydrates, may be used as food, fodder or fertilizer or refined to methane, hydrogen or electricity.

Overconsumption of fossil fuels is self-reinforcing, not self correcting. Consumers want more and bigger cars and consumers want more fuel intensive foods. Food crops need progressively more fuel expensive fertilizers, pesticides and herbicides. Yet there is nothing inevitable about fossil fuel dependence. A change in strategy to sustainable green solar and other carbon neutral renewable energy sources can end dependence on fossil fuels.

Solutions to mega-challenges such as ending the need to burn fossil fuels will be solved by a solution suite of renewable energy sources such as wind, waves, tides, solar, geothermal, nuclear energy and biofuels such as cellulosic ethanol, methane and hydrogen. Algae will play a major role by providing liquid transportation fuels, food, pollution solutions and valuable coproducts.

Viable solutions to these tough problems are only a few years in our future. Producers in the U.S., Mexico, China, Canada and South Africa expect to be growing millions of tons of algae within five years.[1] Makoto Watanabe, a professor at Tsukuba University in Japan believes he can grow algal oil that produces enough biofuel to meet Japan's total transportation needs within five years.[2]

### *Algae*

Nano-sized, single-celled algae are among Earth's earliest life forms. They have been surviving in many of Earth's harshest environments for several billion years. Algae's simplicity enables these plants to be incredibly robust – they not only survive but produce high-value biomass in tough environments. In good cultivation conditions, algae produce protein and energy biomass at speeds that are 30 to 100 times faster than land plants.

Algae are critical to life on Earth as they produce the organic matter at the base of the food chain. The biomass is eaten by everything from the tiniest shrimp to the great blue whales. Algae also produce most of the oxygen for other aquatic life and provide more oxygen to the atmosphere than all the forests and fields combined.[3]

Algae, the Latin name for seaweed, present themselves in all shapes and sizes. Microalgae are single-celled, microscopic organisms often smaller than 25 microns wide. Seaweeds are larger algal species that live in marine environments such as kelps; brown seaweeds that may grow to 180 feet. In tropical regions, coralline algae help build corals and support the formation of coral reefs and other species live in symbiosis with sponges.

Various algae maximize different components. Some species offer over 60% protein and others 90% carbohydrates. The food product, protein, of some species has little natural smell or taste so the product may take on the characteristics desired such as any smell, color, texture, density or taste.

**Figure 1.1 Algae Composition**

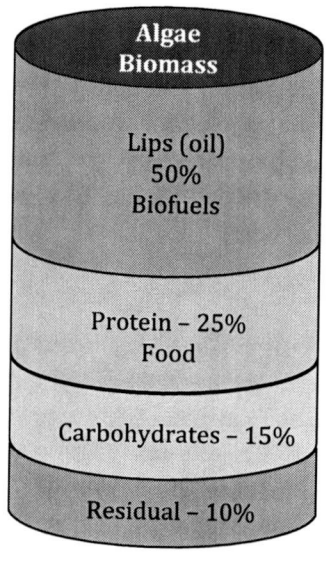

Lipids are substances that dissolve in organic solvents but not in water such as phospholipids, sterol, waxe, chlorophyll, galactolipids, carotenoids and triacylglycerol.

A 20 pound algal biomass with 10 pounds of lipids may produce nine pounds of fuel oil due to gums and ash that are refined out of the lipids to make clean oil for biofuels, foods or cosmetics.

Remaining biomass, coproduct, includes protein, carbs, residual moisture and ash.

Algae are very efficient at converting light, water and carbon into biomass containing oily compounds called lipids that may be extracted and processed into jet fuel, green diesel or gasoline. The remaining biomass, mostly protein and carbohydrate, may be made into foods, medicines, vaccines, minerals, animal feed, fertilizers, pigments, salad dressings, ice cream, puddings, laxatives and skin creams. An example algae composition in Figure 1.1 shows an algal species where 50% of the plant biomass is oil.

Fat algae, also called oleaginous algae, are species that produce large quantities of lipids. Green algae may not look like a biocrude oil feedstock but the petroleum used in today's vehicles is derived from prehistoric biomass which came largely from algae blooms in ancient wetlands and oceans.[4]

Nature's biomass decomposition began over 200 million years ago in the Carboniferous Period under conditions of enormous heat and pressure. Oil pumped from the North Sea consists of decomposed haptophyte algae called coccolithophorids. Algae also make up the major components of diatomaceous Earth, coal shale and coal. The Egyptians built their pyramids with limestone formed from algae.

*Biofuels*

Biofuels have been used since prehistoric times to cook food and to provide heat. Drying cow manure for cooking fires is still a first job for young brides in rural India. Over half the people on Earth use firewood or agricultural materials for cooking and heat. Burning manure robs the land of fertilizer and gathering firewood denudes forests and leads to deforestation.

Biofuels are simply a form of solar energy. Similar to land plants, algae use photosynthesis to convert solar energy into chemical energy stored in the form of oils, carbohydrates and proteins.

Plants used to create transportation biofuels today, primarily corn, soybeans and sugar cane, were domesticated for over ten millennia to maximize food value. They are a convenient but naïve choice as a biofuel because they are critically unproductive in producing energy while heavily resource intensive to grow in terms of cropland, water, fertilizers and fossil fuels.

The two primary types of biofuels are ethanol and biodiesel. Feedstocks for ethanol must be fermented with fossil-fuel heat and use sugarcane and grains, largely corn. Ethanol, an alcohol, can replace gasoline but it requires specially adapted motors because the alcohol dissolves the rubber lines and gaskets.

Unlike ethanol, biodiesel is a clean-burning fuel derived from the vegetable oils of plants such as canola, soy, oil palm, jatropha and algae as well as from animal fat. These oils can be burned directly in diesel engines without engine modification.

Some communities are orchestrating the systematic collection of used restaurant cooking oils and are home-brewing biodiesel.[5] Arizona's Desert Biofuels Initiative's "Gold to Green" project hopes to refine every drop of used restaurant cooking oil in Arizona to green diesel and remove 100 tons of fossil fuel pollutants from Arizona's air each year.[6]

Corn ethanol and soybean biodiesel are considered generation one biofuels because they are extremely inefficient in terms of energy yield per acre and they disrupt crop production. The second generation, which is still in the R&D phase, consists of cellulosic fuels from forest products and dense grasses such as switchgrass which are still inefficient but provide less severe competition with food crops and less water pollution. Second generation biodiesel is also likely to come from algae.

Algal oil behaves similar to vegetable oils because they are essentially the same. When the oil is pressed out of the algal biomass, the oils can be burned directly in diesel engines. It is called clean or green diesel because it burns with almost no pollutants. Alternatively, the oils may be refined into a wide variety of other liquid transportation fuels. The remaining biomass contains carbohydrates which can be fermented and refined with additional heat to produce ethanol.

Biodiesel offers several advantages over ethanol besides avoiding engine modification or redesign. Biodiesel yields about 30% more energy than gasoline and runs much cleaner. In 2000, biodiesel was the only alternative fuel in the U.S. to have successfully completed the

What is Green Algae Strategy?

EPA required Tier I and Tier II health effects testing under the Clean Air Act. A DOE study showed that the production and use of biodiesel, compared to petroleum diesel, resulted in a 79% reduction in $CO_2$ emissions. Hence, biodiesel is often called green diesel.

Algae makes oil naturally and can be refined to make biocrude, the renewable equivalent of petroleum and refined to make gasoline, diesel, jet fuel and chemical feedstocks for plastics and drugs. Algal oil can be processed at existing oil refineries to make just about anything that can be made from crude petroleum.[7]

Alternatively, algal strains that produce more carbohydrates and less oil can be processed and fermented to make ethanol or butanol. Some algal strains contain 90% carbohydrates. The downside of carbohydrate conversion is that current technologies require considerable heat and energy to produce the ethanol. The leftover protein may be used for animal feed or nitrogen rich organic fertilizer.

Some algae create 60% protein and provide additional nutritional elements when compared to conventional land plants. Besides protein, algae contain a broad spectrum of other nutritious compounds including peptides, carbohydrates, lipids, vitamins, pigments, minerals and other valuable trace elements.

Global production and processing of photoautotrophic algae for all species was estimated in 2004 to be about 10,000 tons a year.[8] Other sources suggest that China alone produces and consumes over 100 million pounds of fresh and dried seaweed a year.[9]

Algal protein biomass contains nucleic acids, amines, glucosamides and cell wall materials which may diminish 60% protein content to 50% recoverable protein. Algal proteins are composed of a wide variety of amino acids. The nutritional quality of algal protein is determined by its content and which amino acids are present.[10]

The cellulosic cell wall represents about 8% of the algal dry matter and presents a serious problem in accessing algal protein since it is not digestible for humans and other non-ruminants. Cell walls can be softened by mechanical pressure or heat but those solutions add cost and heat sometimes damages nutritional elements. Several laboratories are working on biotechnology solutions for the cell wall issue. When cell walls become digestible, algae offers a food source that mimics soy protein and may be substituted for any land-based food grains such as wheat, barley, corn, rye, rice or soybeans.

In addition to food, algae provide a wide variety of medicines, vitamins, vaccines, nutraceuticals and other nutrients that may be unavailable or too expensive to produce with land plants or animals. Algae produce the Omega-3 and Omega-6 fatty acids found in fish oils that have been shown to have positive medical benefits to humans.

The carbohydrate component includes valuable cell wall material that is commonly used as thickeners, emulsifiers and valuable pigments. The cellulosic component of some species may be made into paper, textiles or building materials.

Scientists have identified over 100,000 algal and cyanobacteria (blue-green algae) species and there may be several million species. Algae display an infinite number of strains within each species and each may exhibit different characteristics. The U.S. Algal Collection is contains 300,000 herbarium specimens.

***Carbon dioxide sequestration***

An attractive attribute is algae's ability to consume large amounts of $CO_2$, build biomass with the carbon and return oxygen to the atmosphere. Each ton of algae absorbs roughly two tons of $CO_2$. Although the $CO_2$ taken in may later be released when the fuel is burned in vehicles, the $CO_2$ would have entered the atmosphere directly. Reusing $CO_2$ to create renewable liquid fuels makes it possible to prevent the release of $CO_2$ from fossil fuels, thereby decreasing total emissions.

Algae's voracious appetite for $CO_2$ makes it wise to co-locate green solar production systems near coal fired power, manufacturing, breweries or other $CO_2$ source. Water typically contains only about 0.5% dissolved $CO_2$ which enables algae to grow in natural settings. Added $CO_2$ speeds up algae growth by a factor of five or more.

## What is Green Algae Strategy?

In the oceans, algae capture $CO_2$ from the air and from dissolved gas in the water. Algae grow near the surface where sunshine enables photosynthesis. Some algae are eaten by higher members of the food chain which sequesters the $CO_2$. Much of the remaining biomass sinks into the water column where the $CO_2$ is sequestered in the plant bonds. The carbon sink effect is how algae changed the Earth's atmosphere from $CO_2$ rich to $O_2$ rich.

### *Algae's unrealized potential*

Algae's potential productivity is extraordinary and has been recognized as a potential global food source for over a century. Smart scientists recommended strong R&D for algae as a food source after each of the World Wars. The last wave of strong algae research ended in about 1980 when food and fuels were cheap. Cheap foods and fuels provided no incentive to pursue algaculture.

While these scientists exhibited too much technical exuberance, they were right about algae's potential. Unfortunately, sparse R&D combined with the U.S. commitment to corn as a biofuel have limited algae's achievements to a current status of less than 1% of its full potential, Figure 1.2.

**Figure 1.2 Algae's Productivity Potential**

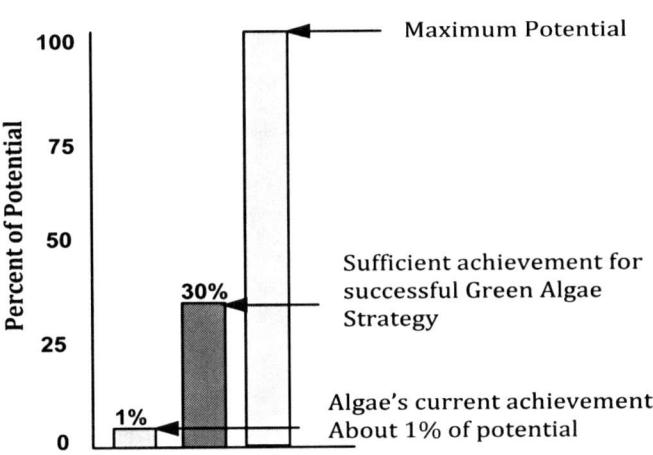

If algae reach only 30% of its potential, the Green Algae Strategy can succeed. Several cultural, political, natural and new technologies have set the stage for accelerated algal industry growth and development.

### What has change since 1980?

In the generation and a half since 1980, food production doubled but at the high cost of using three times as much fossil water for irrigation. Fossil water is mined from deep fossil aquifers that do not replenish with annual rains such as the Ogallala aquifer that supplies most the irrigation water in the Mid-West. Overdraft, the extraction of water over annual rain replacement has been three to 30 times in many crop areas. Overdraft means aquifers will soon crash or water will be too deep to pump economically for food crops.

The doubling of food production during the past two decades brought a false confidence that technology would assure continued production increases. Unfortunately, policy makers not only ignored investments in new food production but allowed the population of many countries to double. The number of starving children and adults has quadrupled reflecting a lack of available food, food prices and food insecurity.

While the need for new food sources has increased, several political factors have changed that benefit algaculture production. Market factors also motivate the search for solutions that decrease food and fuel costs and use less or no fossil fuels.

### Political and Market Changes

**Oil prices.** For decades, the price of a barrel of oil tracked with the price of a bushel of food grain. After the 1973 oil shortage caused by the OAPEC embargo of oil to nations that supported Israel, the price of oil went up dramatically while the cost of food grains stayed flat. (The "A" in OAPEC reflects the embargo participation of only the Arab nations of OPEC.)

Low oil prices provided no incentive for alternative fuel production. Today, concerns about price and fuel security are motivating substantial investments in renewable fuels.

## What is Green Algae Strategy?

**Natural gas prices.** Heavy consumption of natural gas for agricultural inputs such as fertilizer, herbicides and pesticides has led to a 500% increase in prices. Much of the natural gas used in 2008 was not sourced in the U.S. but imported because domestic sources have dwindled.

**Food prices.** For decades, food grains could be produced at about 10% of the cost of growing algae. In the last two years, the price of food grains have doubled and some have tripled and continue to rise. New algal growth models promise to slash production costs.

**Eflation.** Ethanol induced food price increases have escalated the cost of all foods. Burning food for fuel decreases food supply. The U.S. has exported eflation which has led to food riots.

**Food security.** Inexpensive food allowed governments to buy food grains on the world market to feed their hungry populations. When world grain prices double from decreased supply, governments can buy only half as much food.

**Research pendulum.** The U.S. government and governments around the world have recognized the need for truly renewable food and biofuels and algae is the leading contender. Numerous countries, including China, have stopped using biofuel feedstocks that consume cropland because it creates food insecurity.

**Biowar I.** The book *Biowar I: Why Battles over Food and Fuel Lead to World Hunger* details the failure of the U.S. ethanol program and its catastrophic consequences.[11] Burning food for fuel has caused global food riots as predicted and threatens further world food destabilization. The U.S. cannot afford to fund both a destructive ethanol program and sufficient R&D on truly renewable fuels.

**Presidential policy.** The energy and food crisis have motivated every presidential candidate in 2008 to have a strong renewable energy policy.

> **Consumer interest.** Increasing food and fuel prices, water scarcity and pollution are motivating consumers to go green. Issues such as sustainability, renewable energy, carbon, water and ecological footprints are rising toward top of mind for consumers. Global food riots combined with climate change will make R&D for truly renewable foods and liquid transportation fuels mission critical.

Traditional agriculture over consumes its vital non-renewable inputs, especially cropland, fossil water and fossil fuels. Political, social, business and agricultural leaders will examine food sustainable production alternatives that promise sparse natural resource consumption and cause less ecological destruction.

During the last generation while populations were expanding, food production began to hit a ceiling caused by over consumption and pollution of water and extensive use of fossil fuels in an environment plagued with the effects of global warming.

At the top of the list of natural changes is water. Irrigating crops makes sense when water is available near the surface. Food production stops when water tables drop and fossil aquifers crash.

### Nature Changes

> **Water scarcity.** Many communities and cities are facing water shortages for their citizens while 80% of available water goes for irrigation subsidized to farmers at 2% of what city citizens pay. Cities are buying agricultural water rights which means large sections of croplands will return to prairie or desert.
>
> **Water pollution.** Rural and urban communities have found their well-water polluted from agricultural waste streams, especially nitrogen and pesticides. The EPA reported that 37% of U.S. lakes are unfit for swimming due to run-off pollutants.[12]
>
> **Dead zones.** Agricultural waste streams flow into lakes and oceans and create dead zones where all fish, insects, amphibians and plants die from lack of dissolved oxygen.

## What is Green Algae Strategy?

> **Greenhouse gases.** Each acre of agricultural production adds about 2.25 tons of $CO_2$ to the air.[13] Corn production adds additional nitric oxides which is a worse greenhouse gas than $CO_2$.
>
> **Humidity.** Massive production of corn in the U.S. increases humidity as the plants transpire water creating higher humidly for cities and more severe weather, especially thunderstorms that may spawn tornadoes.
>
> **Fossil fuels.** Heavy consumption of fossil fuels has not only led to price increases but to supply disruptions. Another OPEC boycott or loss of distribution infrastructure from violent storms would cause a food cascade causing millions of people to starve.
>
> **Climate change.** Nearly every year since 1993 has been reported to be in the hottest 20 years on record. Nearly new year creates a new record for the hottest days and the hottest nights. Food crops may wilt under intense heat and significantly diminish production.
>
> **Salt invasion.** Rising ocean levels due to hotter surface temperatures have caused massive sea salt invasions of prime cropland from higher tides and storm surges.
>
> **Natural disasters.** In 2008, a terrible cyclone in Myanmar, Earthquakes in China, famine in Africa, drought in Australia, pest vector in Asia, floods in the U.S. Mid-West and wild fires in California have severely diminished available food and resources for growing food.

The changes in nature associated with growing crops in a period of climate change put the entire food production system in jeopardy. Water scarcity represents the most critical constraint. New genetically modified seeds are more productive but consume more water.

While political and nature changes increase the obstacles to increasing production with traditional agriculture, technology breakthroughs improve the landscape for algaculture.

## Technology Changes

**Biotechnology.** New genomic and proteomic technologies make it much easier to understand the mechanisms involved in algal-oil production. One of the challenges researchers have faced is that algae can produce large amounts of oil; as much as 60% of their weight. Unfortunately, they only produce high lipids when they are starved for nutrients. Starving causes them to lose their ability to quickly grow and reproduce.

Breakthroughs in biotechnology will enable researchers to identify paths in the algal genome that control functions such as color, cell wall structure and lipid production. Understanding the molecular switches that increase oil production will be a major finding.

**Automated strain selection.** Researchers have had to perform labor-intensive identification of algal species and characteristics. Automated instruments are 1,000 times more productive than species selection by hand. The same holds true for identification and monitoring algal strains and their characteristics during growth and development.

**Monitoring equipment.** Automated equipment that measures every critical parameter in algae growth enables researchers to perform many times more experiments and tests than were possible previously.

**Nanotechnology.** Nanotechnology enables scientists to understand single-celled organisms that have building blocks the size of nanoparticles – 100 nanometers and smaller.

Some nanoparticle algae are being used to coat thin filaments such as spider web material that can be used in medicine as a platform to grow human cells inside the human body.[14]

**Chemical and mechanical engineering.** Advanced engineering has applied new mathematics and simulations technologies that enable scientists to optimize plant growth, production and

## What is Green Algae Strategy?

> conversion into useful and valuable products. Simulations enable scientists to test hundreds of experiments in virtual production systems which enable faster process optimization.
>
> **Private sector action.** High fuel prices have motivated numerous firms to plan or build commercial production systems. For example, LiveFuels, Inc., is extending the excellent work done by the National Renewable Energy Laboratory, (NREL) called the Algal Species Project. Example company initiatives are highlighted in Chapter 8.

These new technologies have benefited algae. Commercial algaculture will go from thousands of tons to millions of tons in the next few years. While most of the planned production will be for biofuels, the result will also include millions of tons of protein and coproducts.

### *Why algae?*

Current feedstocks for biofuels not only consume precious cropland and trillions of gallons of water but they provide a dismally low yield. The higher oil yield land plants do not grow well in the U.S. including cocoa, rapeseed, jatropha and coconut oil.

Scientists believe commercial algal farms can produce 5,000 gallons of oil per acre annually, Figure 1.3. Corn produces only 18 gallons of oil per acre but produces starches that can be fermented to produce 350 gallons of ethanol per acre. Energy production is based on the energy equivalent of a gallon of gasoline. Corn ethanol is a short-chain hydrocarbon that burns without as much heat as gasoline and produces only 64% of the energy of gasoline.

This makes the energy calculation per acre:

350 gallons of ethanol * 0.64 = **224 gasoline equivalent gallons**

Algae lipids are a longer-chain hydrocarbon (think hardwood compared with softwood) that can be made into jet fuel, JP-8 or green diesel and burn 30 to 50% hotter than gasoline. This makes algae's energy calculation:

5000 gallons of algal oil *1.30 = **6,500 gas equivalent gallons**

Green Algae Strategy

**Figure 1.3 Oil Production Potential – Gallons per acre per year[15]**

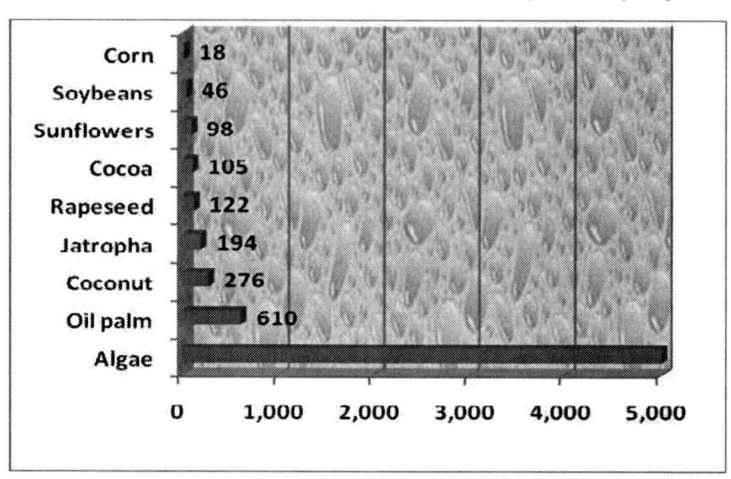

| | |
|---|---|
| Corn | 18 |
| Soybeans | 46 |
| Sunflowers | 98 |
| Cocoa | 105 |
| Rapeseed | 122 |
| Jatropha | 194 |
| Coconut | 276 |
| Oil palm | 610 |
| Algae | |

Algae's potential is about 30 times higher than corn ethanol production. Other parameters such as coproducts, growing requirements and ecological footprint may be even more critical to the choice of changing to sustainable algaculture than oil productivity differences. Algae's potential remains theoretical because scaled production has not yet been achieved. However, significant production breakthroughs are occurring now.

The energy productivity advantage for algaculture occurs largely due to the differences between terrestrial, land-based and water-based plants. Algae express themselves in a nearly limitless number of species and strains which makes them a very unusual organism. Several key characteristics differentiate algae from terrestrial plants.

**Kelp, diatom and fibrous green algae**

 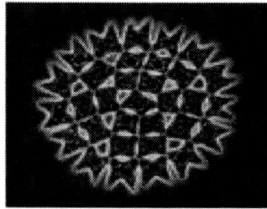 

What is Green Algae Strategy?

*What makes algae special?*

Algae are water-based organisms that grow in fresh, saline, brackish, seawater or wastewater. They learned critical growth, propagation and survival strategies in their several billion of years on Earth. Algae are different from land plants in many ways.

### Algae's Competitive Advantages

**Superstructure.** Land plants invest a large portion of their energy in building cellulosic structure such as roots, trunk, leaves and stems to withstand wind and weather. Algae have no such requirement. Water support algae like a natural womb.

**Growth speed.** Land plants such as food grains require a full growing season from spring to fall – often 140 days or more to produce one crop. Algae learned to flourish when nourished and can grow to maturity and produce over a million offspring in a single day

**Direction.** Land plants grow slowly in one direction, towards the sun and may double their biomass in 10 days and then progressively slow their growth to maturity. Algae grow in all directions, 360°, and may triple its biomass daily.

**Reliable production.** A single event during an entire growing season such as drought, insects, wind or hail can devastate a food grain crop. When bad weather occurs, algae take a rest and slow down their growth rate. When the weather improves, algae immediately continue their explosive growth.

**Composition.** Land plant green biomass such as corn may be 97% non-oil or waste because most of the plant composition is cellulosic structure rather than protein for food or energy producing oils. Some strains of algae produce over 60% lipids – oils that can be converted directly to jet fuels or green diesel.

**Stored energy.** Land plants such as corn can be converted to ethanol that burns with less heat and provides only 64% of the MPG of

> gasoline. Algae convert sunshine, $CO_2$ and other nutrients to long carbon chains that can be converted to more powerful liquid transportation fuels such as JP-8, jet fuel and green diesel that may have 30 to 50% more energy per gallon than gasoline.

The extraordinary potential for algae's commercial production is currently held at bay by difficult yet solvable challenges.

### Algal production R&D

Elements of algal production shown in Figure 1.4 include growing systems, inputs, processing and marketing discussed in Chapter 6.

**Figure 1.4 Algal Production R&D**

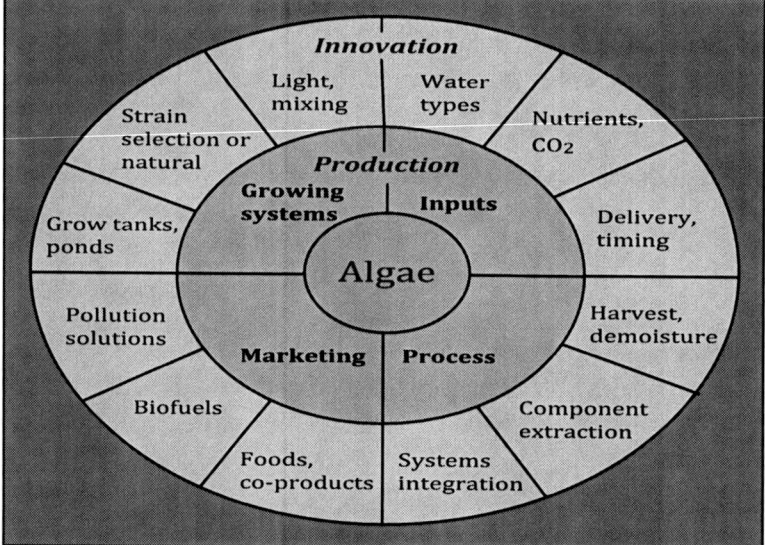

The elements of algal production represent the obstacles to success for the Green Algae Strategy. Fortunately, progress is being made in every step of algal production. Currently, large-scale production systems are carrying the industry because there will be fortunes made in renewable algal biofuels. Micro-scale production for food, cooking fire fuel and fodder are receiving practically no research.

What is Green Algae Strategy?

Marketing may be the weakest link because few algal products have been available to consumers except as food ingredients or health foods. Food manufacturers have used specific strains of algae as subordinate ingredients such as emulsifiers, thickeners and emollients. With the exception of Asia, consumption of algal foods directly has been sparse. Indigenous people in Africa, South America and Mexico consume small quantities of natural stand algae mostly for the vitamins and nutrients they provide.

### Algae Production in a Laboratory

Algae's special characteristics position this plant as a major supplier of globally critical products and pollution solutions; especially for water.

*Algal products and solutions*

Algal products and solutions, summarized in Figure 1.5, include food, biofuels; value added products and pollution solutions. These products and solutions are examined in Chapter 7.

Most companies today are focused on the value proposition for growing algae as a biofuel because the resulting product has so much commercial benefit, especially green diesel, jet fuel and hydrogen.

After the oil is extracted, the remaining biomass may contain 30% protein usable for food. It is also possible that the remaining biomass may contain more value than the extracted oils for components such as pigments, medicines, vaccines or nutraceuticals.

## Figure 1.5 Algal Products and Solutions R&D

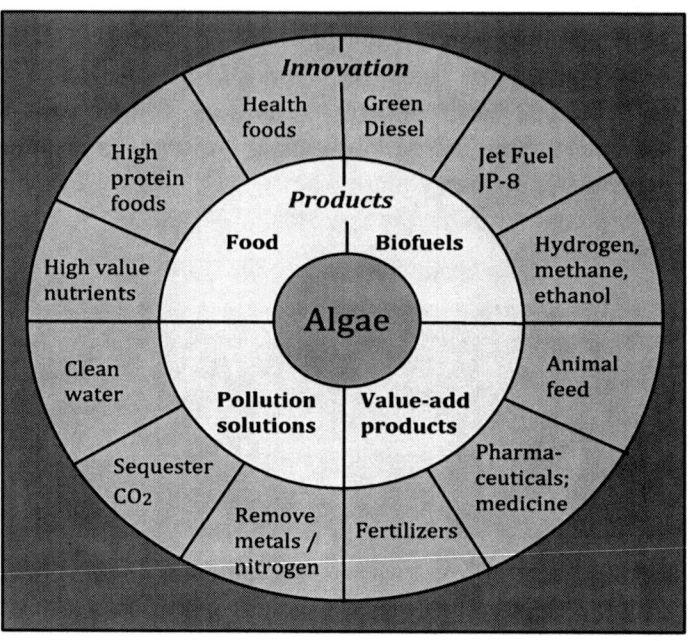

Algae have been harvested for food for more than 4000 years in Africa, Central and South America and Asia. Fossil records and carvings show the Aztecs used algae as a high-protein food.[16] Many indigenous people in the world harvest algae and either eat it directly, use it to flavor soups and stews or feed it to their livestock. Over twenty algae varieties are consumed in Japan.

### *Why have algae remained undiscovered?*

The most common algae question is:

> If algae have so much potential, why has less than 1% of its potential been realized?

For decades, food and fuels were so cheap that there were no incentives for algaculture. Today, soy protein can be grown at about one tenth the cost of algal protein. Fossil fuels can be extracted and refined for about one fifth the cost of algal oils currently. Of course, those numbers will flip with scaled commercial algaculture.

What is Green Algae Strategy?

The rest of the story may be found in political budget decisions, made by both political parties:

- Algae and other truly renewable biofuel feedstocks have lost every political biowar battle to corn. The U.S. government committed to corn ethanol in the 1990s as America's "renewable" biofuel which eliminated federal funding for algae at government agencies like DOE and USDA and University Research Labs.
- Algae have no political action committees or lobbyists.
- Algae have negative sex appeal – NASA receives $17.3 billion for space exploration but produces neither an ounce of food nor a drop of fuel.
- Algae receive no subsidies similar to corn and Big Oil.
- Algae receive no refining subsidy similar to the $0.51 a gallon for corn ethanol.
- Corn ethanol refineries can get bank loans on favorable terms because the business model has been in place for decades and government subsidies modify the risk. Algae refiners cannot get similar private funding because there are no subsidies and no government support to moderate risk.

The U.S. government and the National Renewable Energy Laboratory, NREL, dropped all algae R&D for the last decade to focus on the political choice; corn ethanol. Government grants to universities and independent labs completely evaporated.

The 2007 Energy Security and Independence Act includes language promoting the use of other renewables such as algae for biofuels. Algae began receiving a minute funding in 2007 and NREL has reestablished research on algae. However, algal research still receives a fraction of 1% of the subsidies for corn ethanol.

Algae have remained undiscovered also due to the strong negative social attribution. On an algal belief survey, over 92% of consumers responded with "dislike intensely." They dislike algae because they associate it with icky, stinky green slime.[17] Consumers seem to have a natural aversion toward something they cannot see because it is too small. Of course, consumers cannot see plant cells either but they are familiar with the form of traditional land plants.

Most consumers have near zero knowledge of algae yet they share a very strong negative perception. Consumers are unlikely to embrace a food with a green smelly slime legacy. Algal marketers will have to shift perceptions before consumers will consider adopting algal foods.

A broad set of private equity firms are making some investment in algae but the risks for the first investors in this infant industry are very high. Breakthroughs are likely to enable cheap solar to produce electricity stored in batteries to power cars at an order of magnitude more cheaply than any biofuel. The same probability holds true for wind, waves and geothermal.

However, consumers are going to need liquid transportation fuels for several decades because nearly all the world's existing cars have gasoline engines. The average car produced in Detroit has a road life of about 17 years and will continue to need conventional fuel. Air transportation will continue to be a major consumer of high-value liquid transportation fuels for several decades.

While government funding has been zero, venture capital dollars totaled $84 million in companies developing algae-based fuel in 2008, up from $29 million in 2007, according to the CleanTech Group.[18] These venture capital dollars are miniscule since a simple ethanol refinery costs $250 million to build. Government sponsored health research was $30 billion in 2007.

The serious threats from lack of government funding are that solutions will occur too late to halt climate change causes food crops to fail and a food cascade leads mass migration and the starvation of millions.

Failing public funding, private firms will lock up the fundamental pathways for algae production with intellectual property protections. This is already happening. The world could have viable solutions for green independence from fossil fuels and an end to hunger but the people most in need would not have access.

What is Green Algae Strategy?

America cannot afford to wait. Even though Americans make up only 4% of the global population, the U.S. consume 25% of the world's fossil energy. America has only 3% of the world's oil reserves and currently imports about 65% of the 23 million barrels a day consumed. Imported oil over the next decade will cost more than the entire current national debt, over $10 trillion.

Planting 100% of U.S. cropland used for crops, 440 million acres, in corn would displace less than 12% of oil imports.[19] Using conservative assumptions such as a 30% oil algae, algal oil production could replace 100% of U.S. oil imports on 13 million acres or 3% of American cropland.[20] The National Renewable Energy Lab, NREL, estimates that 10 million acres, about the size of Maryland, as the cropland that could supply 100% of energy imports. The estimate precision is not critical. The clear message is that algae are far more productive as a biofuel than corn or other land crops.

Green Algae Strategy offers complete oil independence for the same total cost as the already failed policy to produce corn ethanol. Green Algae Strategy will save rather than pollute the environment, sequester not emit millions of tons of $CO_2$. Algae production offers additional advantages including that it uses no cropland, no freshwater and no fossil fuels. Therefore, algae does not interfere with the production of traditional food crops, pollute well water and create dead zones or use up dwindling supplies of fossil fuels.

### *What is the Green Algae Strategy?*

Green Algae Strategy positions algae as central to solving the most critical problems facing human societies today including restoring the health of our planet:

- Independence from oil imports and halt use of fossil fuels
- Reverse global warming by sequestering CO2
- End hunger in America and the world
- Stop smoke death caused by cooking fires

The Green Algae Strategy map outlines the critical actions. How Green Independence is accomplished is described in Chapter 10.

### Figure 1.6 Green Algae Strategy Road Map

| Actions | Description | Timeline |
|---|---|---|
| 1. Announce Green Independence | • End U.S. oil imports<br>• End U.S. use of fossil fuels<br>• End fossil fuels globally<br>• Recapture carbon released in fossil fuels<br>• End hunger in America<br>• End hunger globally<br>• End smoke death | 3600 days<br>20 years<br>40 years<br><br>75 years<br>10 years<br>30 years<br>20 years |
| 2. End Biowar I | End ecological destructive subsidies such as corn, water, power, ethanol and Big Oil. | |
| 3. Appoint a Cabinet level Chief Sustainability Office | Create a sustainable food and fuels policy that protects critical U.S resources including air, soils, water and other natural resources and the environment. | |
| 4. R3D for large and small-scale biofactories that enable local production | Invest $50 billion a year in all forms of carbon neutral, sustainable clean energy.<br><br>Invest $10 billion a year in algal research, development, demonstration and diffusion focused on global technology transfer that makes local algal production possible. | |
| 5. Encourage green innovation and collaboration | Create collaborative social networks that use open source technology and solutions available everyone on Earth. Engage venture capital firms, universities, science centers, churches and communities to pursue green and sustainable activities. | |
| 6. Shift disposable consumption to sustainable | Engage and educate consumers in critical consumption decisions such as sustainable food, transportation and lifestyles. | |

What is Green Algae Strategy?

| | |
|---|---|
| 7. Create and monitor green metrics | Create green metrics and use both government and NGO's and universities to monitor and report progress, concerns and insights. |
| 8. Communicate and celebrate. | Recognize people, innovations and progress toward sustainability and green independence goals. |

Announcing Green Independence with goals and timetables will create the motivation and focus that will enable achievement. Successful solutions can occur quickly only when subsidies for ecologically destructive and non-sustainable agriculture and fossil fuels are ended and those resources shifted to renewable and sustainable production of food and energy.

America desperately needs a sustainable foods and fuels policy. A sustainable foods and fuels policy will require trade-offs but should not sacrifice the next generation's groundwater for a biofuel. The policy should integrate the relevant government agencies in order to avoid the catastrophic failures associated with ethanol.

Current energy policy sacrifices precious land and non-renewable water for a weak fuel additive. The U.S. government spends billions on national security while ignoring water and ecological security. The American people would be well-served by a Cabinet level Chief Sustainability Office that would orchestrate sustainable food, fuels, transportation, health and environment. The merger of the National Oceanic and Atmospheric Administration, NOAA, and the U.S. Geological Survey, USGS as recommended by their former administrators, D. James Baker and Charles Groat respectively seems a logical place to start.[21]

Research and development needs to broaden to demonstration and diffusion and include the needs of developing countries.[22] Small-scale algaculture production would most benefit developing countries so local labor can provide sustainable production.

Knowledge works most powerfully when it is widely shared. Green solar knowledge provides a common base of understanding, action and technology development. A global understanding of how to grow food and fuels locally will give millions the opportunity to provide for their family and their community.

Collaborative networks need to be built for technology transfer and the maximization of open source solutions that benefit all people on Earth. Collaborative networks may monitor the algal industry to avoid the unintended consequences that undermined corn ethanol. Breakthrough technologies typically spawn some unintended consequences. Anticipating, monitoring and reporting issues and outcomes can mitigate social, economic and ecological damage.

A Chief Sustainability Office could work with psychologists and consumer behavior professionals who know the science of influence. They could lead consumer education, create environmentally sensitive product labeling and incentives for making sustainable consumption choices. Disincentives might be considered such as luxury taxes for conspicuous overconsumption for such precious resources as water, energy, pollution and waste products.

Successful innovators need to be recognized and rewarded for their contributions towards sustainability. Similarly, communities and institutions might create green metrics such as waste avoidance, water saved and energy saved. Then communities can track and celebrate progress toward sustainable goals.

Achieving the Green Algae Strategy will not be easy because there are obstacles to successful algal production. If achievement were easy, it would already have been done and that would eliminate the fun.

A global environmental scan indicates an urgent need for food, water and energy solutions. We live on a small planet where people today are very hungry, thirsty and needy.

# Chapter 2. What are the Global Challenges?

> Every gun that is made, every warship launched, every rocket fired signifies, in the final sense, a theft from those who hunger and are not fed, those who are cold and are not clothed. This world in arms is not spending money alone. It is spending the sweat of its laborers, the genius of its scientists, the hopes of its children . . . Under the cloud of threatening war; it is humanity hanging from a cross of iron."
>
> -Dwight D Eisenhower

Every society on Earth must choose among a variety of public policy options where public funds support food, agriculture, education, national defense, transportation, pollution and health.

Three key global challenges addressed here are hunger, biofuels and water. Each issue causes significant difficulty for America as well as most people and governments on our planet. The humble algae plant provides potential solutions in each area. Possibly the most critical challenge is food security because so many people live on the edge of starvation and climate change threatens traditional agriculture.

Green Algae Strategy

*World hunger*

The academies of sciences from 58 countries, including the U.S. National Academy of Sciences, summed up the global situation in the *Population Summit* report:

> Humanity is approaching a crisis point with respect to the interlocking issues of population, food, natural resources and sustainability.[23]

The report documents the serious status of vital resources needed to support human life. Similar to global warming, science has not computed the carrying capacity of Earth but the report makes clear that current practices are not sustainable. The report recommends local solutions give way to more cooperative global solutions to support food security and human life. Food security means access by all people at all times to sufficient food. Yet a hunger death occurs every 3.6 seconds.[24]

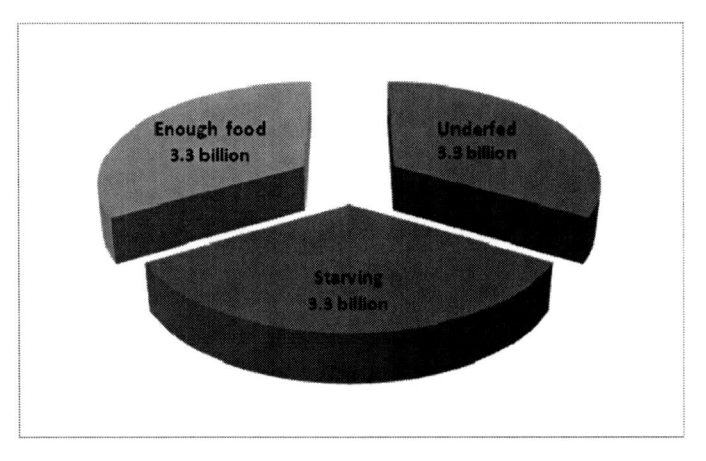

**Figure 1.1 Status of World Hunger**

- Enough food 3.3 billion
- Underfed 3.3 billion
- Starving 3.3 billion

The U.N. Food and Agriculture Organization, FAO, estimates that 850 million people are chronically undernourished. Undernourished mothers give birth to low birth weight babies who are stunted and usually exhibit mental and physical impairments. Currently in India and Bangladesh, one in three babies begin their lives with severely low birth weight. Undernourishment and stunting frequently overlap

## Global Challenges

with the vitamin and mineral deficiencies that affect nearly 2 billion people worldwide.[25]

Jacques Diouf, FAO Director General, presided over the 1996 World Food Summit in Rome that proposed to reduce by half the number of undernourished people by 2015. Dr. Diouf recently announced failure in that goal and "far from decreasing, the number of hungry people in the world is *increasing* – at the rate of 4 million a year."[26]

Jacques Diouf describes poverty in an unusual word picture:

> If poverty could be photographed, it would show a family of landless peasants – the poorest of the world's poor. Coming second to them in this cheerless classification are the people with plots of land so small and depleted that they cannot produce enough to feed themselves. The value of this picture is the clarity of its message: land – or to be more precise, the lack of it – is one of the root causes of world hunger and poverty.[27]

Jeffry Sachs, Director of the Earth Institute at Columbia University and author of *The End of Poverty: Economic Possibilities for Our Time* calls this the poverty trap where peasant farmers are caught in the spiral of rising population with stagnant or diminishing food production per person. Unfortunately, more people are falling into the poverty trap due to a lack of cropland, water shortages and fertilizers and from global warming that spawns severe storms and expands deserts.

The U.N. Report on Biofuels, May 2007, said:

> Liquid biofuel production could threaten the availability of adequate food supplies by diverting land and other productive resources away from food crops. Many biofuel crops require the best land, lots of water and environment-damaging chemical fertilizers.[28]

Loss of food security – scarcity and price – will hit the most vulnerable 2.2 billion people who are already starving because they cannot afford the price of food today for themselves or their family. Over 2 billion people subsist on less than $2 a day. Food aid programs typically have fixed budgets and if the price of grain doubles, food aid is reduced by half. A small increase in food prices or lack of supply will cause many millions to die from malnutrition and associated diseases.

In 2008, America will spend over $20 billion on subsidies to burn nearly 100 million tons of corn for ethanol and displace less than 3% of foreign oil imports. It seems difficult to rationalize a policy that burns food for a weak fuel additive when 30 million children under five die from starvation every year.[29]

Over 60 million Americans received federal food assistance in 2006 under the $53 billion USDA food assistance programs which include Food Stamps, National School Lunch, Special Supplemental Nutrition Program for Women, Infants and Children and the School Breakfast Program. Food assistance expenditures have increased each of the past six years and annually set historical records.[30] The rising cost of corn raises all food prices which significantly adds to the cost of food assistance.

The working poor make extraordinary sacrifices but receive little in return. The USDA charged with supplying food for the nation's poor, spends half its budget on food stamps and lunch programs.[31] Nearly 31 million Americans, 1 in 10 U.S. citizens, receive food stamps. Over half of the recipients are children and 8% are over age 60. Food stamp recipients are impacted by ethanol because burning food causes eflation, ethanol induced prices increases. The price of milk, cheese, chicken, pasta and corn products have increased over 50% due to ethanol so the average $94 monthly food coupon buys less food. When the price of pasta doubles, food stamp users can buy only half a pound. Hungry Americans, especially children, get less food.

While increasing numbers of Americans struggle to survive on USDA food assistance, the USDA pays farmer subsidies and helps orchestrate the ethanol industry. The USDA reports show each year more bushels of corn burned for ethanol. Converted to millions of tons of corn, those figures rise dramatically in 2007 and 2008, Figure 1.5. Where is the logic in a strategy that burns 100 million tons of food while distributing $1-a-meal food stamps to 31 million hungry Americans?[32]

Global Challenges

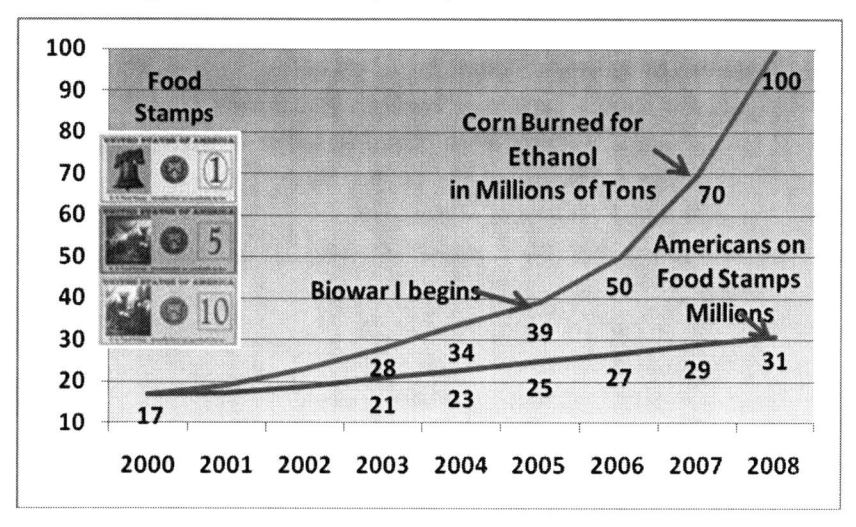

**Figure 1.5 Food Stamp Recipients and Corn Burned**

[Chart showing two lines from 2000-2008:
- Food Stamps (Americans on Food Stamps, Millions): 17, 21, 23, 25, 27, 28, 29, 31
- Corn Burned for Ethanol in Millions of Tons: 34, 39, 50, 70, 100
- Annotation: "Biowar I begins"]

A situational analysis of our planet's environment indicates a critical need for algae to live up to its productivity potential.

*Situational analysis – environment*

Population growth and global warming, water insecurity and pollution make a deadly combination for millions of people on all continents. The world's demand for food is expected to double within the next 50 years, while the natural resources that sustain agriculture will become increasingly scarce, degraded, and vulnerable to the effects of climate change.[33] Water especially has passed its tipping point with global warming as so much water has been degraded, depleted or diverted.

In many poor countries, agriculture accounts for half of GDP and 80% of employment. About 70% of the world's poor live in rural areas and most depend on agriculture for their livelihoods. These people are being adversely affected by climate change, especially heat and drought. Consider that we are currently facing these historical records since humans were first on Earth, for the most:

- Food insecure global citizens, exceeding 3.5 billion
- Hungry Americans on food support, more than 60 million
- Deep water tables and aquifer crashes

- Hottest days, years and decades
- $CO_2$ levels in the atmosphere
- Greenhouse gas production
- Fierce hurricanes, cyclones, tornados and storms
- Springs, rivers, lakes, reservoirs and wetlands that go dry
- Cities facing water supply and quality crises
- Polluted rivers, lakes and well-water
- Pollution caused dead zones in rivers, lakes, estuaries and oceans
- High ocean surface temperatures
- High ocean tidal and storm surges
- Severe forest, rangeland and wild fires
- Water bodies unfit for human recreation or use
- Acute human and animal pesticide poisonings

Global warming, ethanol production and population increases are creating historical lows in:

- U.S. grain stores (food stored for an emergency)
- Global grain stores
- Arctic ice sheet coverage
- Rainfall in many areas
- Water levels in reservoirs and lakes
- Aquifer and water table levels
- Glaciers and snow packs
- Coral reefs

At the time all these Earth records were severely impacting world food production and supply, Biowar I began.

### *Biowar I*

In this hungry and thirsty environment, American politicians decided to ignite Biowar I where food or poisonous agents serve as destructive weapons. In Biowar I, food is sacrificed to produce a weak fuel additive.[34] Burning nearly half the production of America's primary food crop as fuel is unsustainable and wastes valuable non-renewable resources, especially fossil energy and water.[35]

Global Challenges

Burning massive amounts of a food crop as fuel destroys domestic food supplies; decimates U.S. food exports and spikes domestic and international food prices causing food riots. Growing corn for ethanol as well as burning ethanol as a fuel adds to greenhouse gasses and causes smog. Corn ethanol production poisons soils, wetlands, rivers, lakes and well water from fertilizers, herbicides and pesticides.

Growing immense amounts of corn represents ecological suicide as it drains trillions of gallons of non-replenishable groundwater putting the next generation at risk of having no groundwater. Groundwater is being mined at a non-sustainable rate to irrigate corn for ethanol.

Biowar I inflicts costs, casualties and catastrophe in a magnitude far greater than a conventional war. Taxpayers are forced to pay over $20 billion annually to subsidize the erosion and pollution of our air and water resources for a tiny replacement of foreign oil. America has insufficient disposable cropland, water or energy to waste on a policy that fails all its objectives. Corn is a poor choice as a biofuel because compared to alternatives: corn:

- Requires more water, land, fertilizer, herbicides and pesticides
- Severely pollutes air, soils, rivers, lakes and well-water
- Degrades and erodes soils
- Grows slowly and produces a low energy biomass yield

The U.S. has 134 ethanol plants with refining capacity of 7.2 billion gallons annually. Another 77 plants are under construction and will add 6.2 billion gallons more capacity for a total of 13.4 billion gallons by the end of 2008.

Since current production averages 350 gallons per acre of corn, ethanol will consume roughly 40 million acres of cropland and commandeer nearly half of the U.S. corn crop while displacing less than 3% of U.S. oil imports. Algae growing on 2 million desert, non-cropland acres could produce more energy and leave the cropland free to grow food.

Corn ethanol is not sustainable because it consumes too many non-renewable resources – water, cropland, fertilizer, energy and subsidies.[36] The direct and indirect costs of the ethanol industry are neither sustainable nor sensible for farmers, consumers or taxpayers.

**Gulf of Mexico Dead Zone**

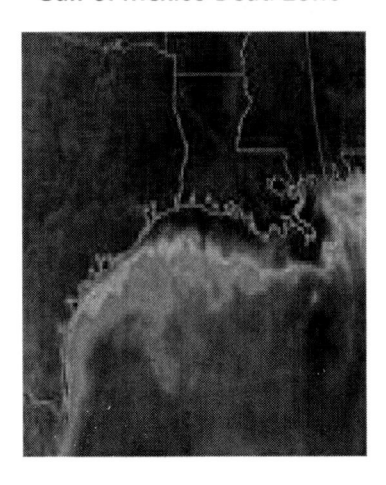

The USDA's research shows that each acre of corn erodes six tons of soil each growing season from wind or water.[37] The resulting dead zone off New Orleans in the Gulf of Mexico has grown to the size of New Jersey. A dead zone occurs from agricultural run-off where all fish, plants and other living things die from lack of oxygen from exploding bacteria populations.

The USGS reported the 2008 spring run-off carried 817,000 tons of nitrogen, roughly 45% above normal, and 85,000 tons of phosphorous about 85% percent above normal to the Gulf.[38]

Corn-producing states tend to have higher pollution rates such as Oklahoma where 94% of lake acres fail federal water quality standards and 75% of the rivers are polluted.[39]

America's current focus on corn ethanol creates a severe negative ecological footprint. Ethanol is promoted as a clean burning fuel but burning ethanol creates smog and the production of corn increases rather than reduces greenhouse gases, especially $CO_2$ and nitric oxide. Ethanol refineries are so highly polluting that the EPA exempted refineries from clean air and clean water laws. Agricultural production is also exempted from many clean air, soils and water laws in support of ethanol production.

A team of over 30 scientists reported in *Science* that over a third of ocean nitrogen is human-sourced pollution, especially from fossil fuels burned in power plants, vehicles and agricultural run-off. The increasing nitrogen pollution is causing havoc in the oceans. The team predicts expanding dead zones and significant changes in species that populate surface areas.[40]

Removing food from food supplies to create fuel causes **eflation**, ethanol induced price increases.[41] Eflation has pushed up prices for not only corn but all agricultural inputs such as diesel oil, natural gas, crop land, farm equipment and especially foods. The World Bank reported that all the extra corn produced from 2004 to 2007 was consumed by ethanol.[42]

The price of rice, corn and wheat has doubled in the past two years and continues to rise.[43] Bandits in multiple countries steal food from stores, warehouses, armed military convoys, silos and fields. NGOs cannot deliver food in many countries because drivers are not safe.

Jean Ziegler, of the United Nations Right to Food Program said the biofuel policy pursued by the U.S. is one of the main causes of the current worldwide food crisis. He called the actions a "criminal path" contributing to an explosive rise in global food prices through using food crops to produce biofuels. He warned of worsening food riots and a "horrifying" increase in deaths by starvation before reforms could take effect.[44]

A nutritionist with the United Nations World Food Program said that "global price rises mean that food is literally being taken out of the mouths of hungry children whose parents can no longer afford to feed them."[45] A World Bank report on rising food prices noted increases in global wheat prices reached 181% over the 36 months leading up to February 2008, and global food prices increased by 83 percent.[46]

Josette Sheeran, Director of the World Food Programme, said in Ethiopia: "The cost of our food has doubled in just the last nine months. We are seeing more urban hunger than ever before because people are unable to afford food."[47]

World Bank president, Robert Zoellick, warned that 33 nations are at risk of social unrest because of the rising prices of food.[48] For countries where food comprises from half to three-quarters of consumption, there is no margin for survival.

The U.S. has exported eflation by burning food for fuel which lowered world food supplies, increased prices and ignited severe food riots globally. Vietnam, Cambodia, Egypt, India, the Philippines and Thailand have stopped crop exports. Philippine leaders warned that people hoarding rice could face sabotage charges. China has also threatened criminal charges for hoarders. A moratorium is being considered on using agricultural land for housing or golf courses. Fast-food outlets are being pressed to offer half-portions of rice.[49]

Food riots have recently broken out in Mexico, India, Morocco, Egypt, China, Pakistan, Zimbabwe, Namibia, Italy, Austria, Hungary, Uzbekistan, Indonesia, Yemen, Guinea, Cameroon, Burkina Faso, Mauritania and Senegal. Examples in 2008 include:

- **Bangladesh.** Families spend up to 70% of income on food and more than 50,000 households are getting emergency food after rising rice prices.
- **Cameroon.** At least 24 people killed and 1,600 people arrested in February. Taxes slashed on food imports and public sector wages increased by 15%.
- **Indonesia.** Over 10,000 demonstrated outside the presidential palace in Jakarta after soybean prices rose more than 50% in a month and more than 125% over the past year.
- **Egypt.** Seven people have died in fights or of exhaustion queuing for subsidized bread. Dairy products are up 20%, oil 40%.
- **Burkina Faso** in West Africa. Riots in three towns after the government promised to control the price of food but failed.
- **Guinea.** Multiple anti-government riots over escalating increases in food prices in the past 18 months.
- **Pakistan.** Thousands of troops have been deployed to guard trucks carrying wheat and flour.
- **Haiti.** Protests at high cost of living descended into riots and four people were killed in clashes with security forces.
- **Mexico.** Over 75,000 people protested the run-up in the price of corn tortillas to as much as doubled or tripled in a year.
- **Yemen.** Tanks were deployed in Yemen after five days of angry protests by thousands of mostly young people who blocked

roads and torched police stations and military vehicles. At least 14 people were killed as the food riots spread to multiple cities and people protested wheat prices that had doubled in March.

Robert Zoellick, stated at the 2008 Conference on World Food Security in Rome: "Riots in over 30 countries, 30 million Africans who will likely fall into poverty, 100 million people worldwide who are at risk, 850 million people who are malnourished; two billion people who are struggling every day to put food on the table. If we cannot act now, when?"[50]

How could people in hungry countries not blame the U.S. for food shortages and price increases when prior to the ethanol program America provided half the world's foodgrains and 70% of the world's corn imports? How could Mexico not blame the U.S. when American corn subsidies displaced over a 1.5 million poor Mexican farmers?[51] Farmers were forced to leave their land because they could not compete with subsidized U.S. corn. Many of these farmers added their feet to the flow of illegal immigrants to the U.S. from Mexico.

British International Development Secretary Douglas Alexander said "It's unacceptable that rich countries still subsidize farming at $1 billion a day, costing poor farmers in developing countries $100 billion a year in lost income."[52]

A group of more than 400 agricultural experts, known as the International Assessment of Agricultural Knowledge, Science and Technology for Development concluded through its global and regional studies report that governments and industries need to discontinue environmentally damaging farming methods. At their 2008 meeting in Johannesburg South Africa, the group recommended "ending subsidies that encourage unsustainable practices."[53]

Consider a case study where the 2008 U.S. production of 10 billion gallons of ethanol is replaced in possibly 2012 with algaculture biofuel.

## Table 1.1 Corn Ethanol versus Algaculture for Biofuel

| Production: 10 B gallons | Corn ethanol 2008 | Algaculture biofuel Possibly 2012 |
|---|---|---|
| Gasoline equivalent energy | 6.4 billion gallons of smoggy ethanol | 13 billion gallons of clean green biodiesel |
| Prime cropland | 40 million acres | 0<br>2 M desert acres |
| Fossil fuels used | 6 billion gallons of dirty diesel | 0 or minimal<br>10 M gallons of biodiesel |
| Fuel production used by farmers | Zero – ethanol is too weak for farmers | 100% – clean, green biodiesel |
| Gallons of fresh irrigation water | 2 trillion | 0 or minimal brine or waste water |
| Pesticides and herbicides | Millions of tons | Almost zero |
| Soil erosion | 240 million tons | None |
| Water pollution | Severe – ag chemical waste stream | None |
| Ecological footprint | Severe – ag chemical wastes | None |
| Sustainable production | Renewable but not sustainable | Yes |

Algaculture would save 40 million acres of prime cropland, 6 billion gallons of fossil diesel, 2 trillion gallons of fresh water and avoid millions of tons of pesticides and herbicides from entering the U.S. ecosystem each year.

Biowar I can end quickly by stopping ecologically destructive subsidies for corn, irrigation water for corn and ethanol refining. If corn ethanol makes sense as a biofuel, let it compete without subsidies.

Removing billions of dollars from ecologically destructive subsidies will free funds to establish truly renewable energy technologies that are far more productive than corn.

American taxpayers are going to ask a simple question:

> If corn ethanol is not sustainable, pollutes our air, soils and water and causes catastrophic unintended consequences, why are taxpayers forced to spend in excess of $20 billion a year on a failed policy?

Green Algae Strategy offers sustainable alternatives for food and biofuels as well as independence from foreign oil.

### *Corn coproducts*

Ethanol supporters argue that the corn coproduct, distillers' grain, adds value but the value is trivial. Left wet, it is likely to grow mold or spontaneously combust – burst into flame from the heat generated from rotting – similar to green hay. To avoid mold or combustion, it must be dried and pelleted, requiring more energy and cost, before it is trucked to feedlots. Distillers' grain has other problems too:

- The concentrated corn mash causes indigestion so it cannot be fed to poultry or hogs except in low doses.
- Biomagnifies (concentrates) toxins present such as aflatoxin which occurs when corn is grown in hot areas like Arizona and western Kansas, Oklahoma and Texas.
- Negatively affects marbling characteristics – the intramuscular fat that makes meat more juicy, tender, and flavorful.
- Has only a 4 day shelf-life wet and then begins to grow mold.

- Is overproduced so refineries in 15 states must pay to first dry it and then ship it to livestock producers in other states.[54]

Iowa State University reported Iowa alone produces about five times more distillers' grain than there are cows in the Midwest to eat it.[55]

### Cellulosic ethanol

Ethanol supporters contend that cellulosic ethanol will overcome corn's low biomass productivity. Cellulosic ethanol is produced from waste wood or weeds such as switch-grass or elephant grass by enzymes or thermo-chemical reactions that break down the cellulose. The cellulosic woody plant structure is turned into sugar which can be refined with fermentation and lots of heat to make ethanol.

Cellulosic ethanol presents several problems:

- In order to grow grasses with enough biomass for biofuel, large areas of cropland will be necessary.
- Grasses are likely to need supplemental fresh water irrigation, especially for germination.
- Much of the marginal lands proposed for grasses are currently protected wetlands and natural habitats
- The cost of refining cellulosic ethanol is 4 to 10 times corn.
- The enzymes that are supposed to break down the structures do a great job until the mixture approaches half the alcohol of wine, 6%, at which time they get alcohol poisoning and die. (Ethanol is 200-proof corn whiskey, so enzymes have a long way to go.)
- Thermo-chemical reactions diminish ethanol's net energy value. (NEV measures how much energy goes in versus energy yield.)

A renewable fuels expert, Al Darzins at the National Renewable Energy Laboratory, has studied cellulosic ethanol intensively and believes practical production is at least 10 years away.[56]

Several ethanol feedstocks that also consume cropland offer much higher productivity than corn. Sugarcane used by Brazil as their ethanol feedstock yields over double the production of ethanol per acre compared with corn. Brazil is blessed with lots of cheap labor and inexpensive flat land fed by rain. Sugarcane does not grow well in the U.S. climate.

Corn ethanol costs about 30 to 50% more than sugarcane ethanol because the corn starch must first be converted to sugar before being distilled into alcohol. A 54-cent per gallon import tariff imposed by the Energy Tax Act of 1978 and renewed in the 2005 Energy Policy Act effectively stops Brazilian biofuel imports. The trade tariff is promoted by the powerful American sugar lobby, which does not want a competitor to high-fructose corn syrup, domestic sugar interests and of course, the farm lobby and ethanol producers.

### *Food solutions*

Growing algae on 13 million non-cropland acres could provide 100% of U.S. imported energy and put 40 million acres of corn ethanol cropland back into production for food crops to feed Americans and the world.[57]

Protein, a coproduct of algae grown as a biofuel could feed millions of people or animals. Production systems optimized for fuel may lack sufficient cleanliness for human foods but the surplus protein could feed millions of animals and supply many tons of organic fertilizer.

Algae cannot be used as a human food today because the cell walls are not digestible. To be used as a human food directly, mechanical solutions, strain selection or bioengineering must solve the cell wall problem. Digestible cell walls will create the tipping point that enables algae to serve the world as food.

Food production depends on water that is increasingly scarce. Water represents the limiting resource for sustainable foods; more vital than cropland or energy. Food fails to grow without abundant water.

### *Water*

> The sage's transformation of the World arises
> from solving the problem of water.    Lao Tze

A solution to the challenge of water, a critical issue throughout the ages, remains possibly the most vital global issue today, especially for food production. War and water, in English, have only two different letters – and may suggest a connection. Failing sufficient water, crops die and a food cascade, which operates like a bank run, threatens to consume people and their communities.

Solutions to the problem of water offer only two alternatives:
1. Find, harvest and transport more water
2. Develop a biofuel source that requires minimal fresh water

Alternative one replicates the unsustainable actions of the last 50 years – using wider pipes, larger pumps and deeper holes to mine more non-renewable water faster to grow crops. Even if pumping energy were free, this approach crashes aquifers and undermines the land for both crops and people.

Alternative two leaves the aquifers in place to support food crops and grows biofuels that use no or minimal fresh water and nominal cropland yet demonstrate high energy productivity.

Green Algae Strategy applies alternative two and grows biofuels in water unsuitable for crops, saving fresh water for food production. Algae grow effectively in salt or wastewater and can clean the water – bioremediation – which can be used for irrigating terrestrial crops.

Lack of land presents a serious predicament because food production requires not just land but good cropland. Even good cropland produces only dust with insufficient water. At the time when the number of hungry people has reached record highs, acute water scarcity has struck countries in the Middle East and North Africa, as well as Mexico, Pakistan, South Africa and large parts of China and India.[58] Diminishing cropland, severe drought, global warming, fierce storms and crashing aquifers combine to make world food supplies precarious.

Worldwide, 70% of all the water diverted from rivers or pumped from underground is used for irrigation.[59] Industry uses 20% and residential users consume 10%. With the demand for water growing steadily in all three sectors, competition is intensifying. In this contest for water, farmers almost always lose to money – cities and industry.[60]

The human body is about 60% water and each of us need about a half a gallon daily for drinking and at least 500 gallons to produce the food for a vegetarian diet.[61] In affluent societies where grain consumptions takes the form of dairy and meat products, the California Farm Bureau estimates daily water consumption exceeds 4500 gallons.[62] Hence, efficient water use is critical for agriculture, industry and consumers.

Water represents the vital limit to growth in food production. The world currently grows nearly twice as much food as a generation ago but extracts three times more water from rivers and especially aquifers to support this production. Powerful modern pumps draw water from deeper and deeper wells – at an unsustainable cost of both power and water. Unsustainable use of water foretells severe hunger or starvation for future generations.

The American economy depends on irrigation which accounts for 81% of water use throughout the U.S.[63] Unfortunately, and despite technical advances over the past 20 years, 50-80% of irrigation water leaks or evaporates before reaching crops.[64] The USDA reports that irrigated cropland accounts for about 50% of total crop sales.[65] Consequently, groundwater for irrigation and drinking represents a critical – but limited and rapidly diminishing – strategic resource.

**Consumptive water use.** A critical issue for sustainable water management focuses on how water is used. Extracted or surface water that cannot be recovered and reused locally is called consumptive use.[66] Irrigation is consumptive because the water evaporates or transpires from crops into the atmosphere creating water vapor. Rising air currents carry the water vapor upward, high into the atmosphere, where the air cools and loses its capacity to support the moisture. The water vapor condenses to form cloud droplets, which may eventually combine with other droplets and produce precipitation, Figure 2.1.

Water vapor arising from irrigation may fall as rain 2,000 miles away but is lost for local consumption. The extracted moisture falls not on cropland but floats on the wind to fall on mountains or oceans. From a farmer's point of view, consumptive water is gone. Lost water does not renew the local area's soil moisture, wetlands, wells or aquifers.

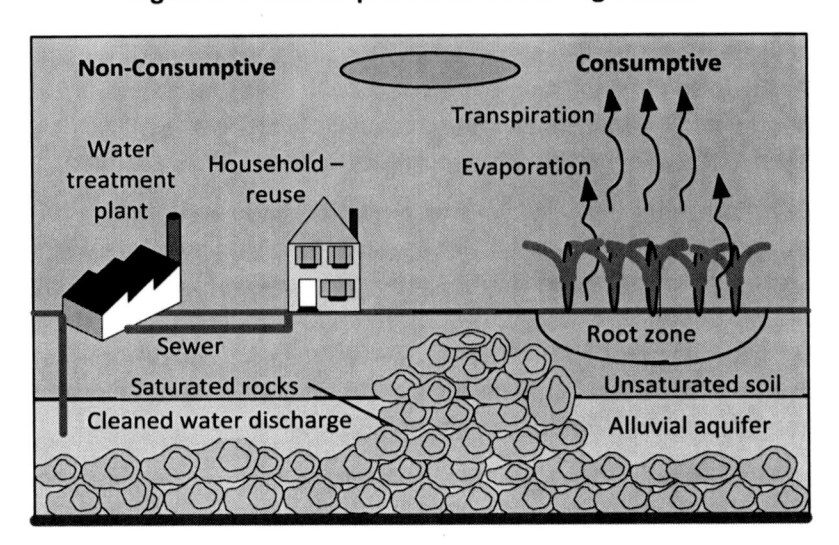

**Figure 2.1. Consumptive Water Use – Agriculture**

Corn and other food grains inefficiently incorporate water into the plant biomass. Corn takes in water through the roots to deliver nutrients to the leaves. The plant then releases most of the water through small pores on the undersides of the leaves called stomates. Corn protects itself from heat with transpiration that acts similar to evaporative cooling to stabilize the plant's temperature.

To get sufficient water, farmers must apply far more water than corn actually incorporates in the biomass. Most of the water is lost from the field as either soil evaporation or plant transpiration. Some irrigation water seeps down below the corn's root zone and also is not available for reuse.

A single acre of corn gives off about 4,000 gallons of water each day from soil evaporation and plant transpiration.[67] People who live near a corn field can feel the extra humidity produced by the escaping water vapor. Water vapor is a greenhouse gas that accelerates global warming as it absorbs and radiates the sun's rays. The high water loss from corn also means more consumptive use water must be provided to sustain the corn field's growth.

Water used outside homes such as on lawns, gardens and in pools as well as city parks and golf courses is similarly consumptive and lost for reuse. Household water for farms or communities that use septic tanks rather than sewers is also consumptive.

City water for household use is **non-consumptive**. Water used in the home goes down the drain and empties through the sewers to the local wastewater treatment plant. The plant removes impurities, possibly with algal screens and dumps the clean water into a nearby stream or water source. Mississippi river water, similar to the waters of other rivers, is reused many times by cities on its way to the Gulf.

Unfortunately, roughly 80% of extracted water goes for consumptive use in agriculture.[68] The conservation programs promoted by local water companies for household water use such as low flow toilets, faucets and showers have only a trivial impact on water conservation compared with agriculture because household water is reused. Household water conservation programs primarily impact the amount of water flowing to the wastewater treatment plant; not water loss.

Water conservation education, especially about yards and other non-household uses, builds an important sense of community and educate consumers about sustainable water management. Consumers who conscientiously conserve their community's water are prepared to convey their sustainability concerns to politicians who currently support ecologically damaging subsidies such as irrigation that waste water.

**Aquifer depletion.** Over 65% of U.S. irrigation extracts water from underground aquifers which are composed of sand, gravel and other materials with gaps large enough to hold and transmit water.[69] Aquifers display all the variability associated with surface geologic formations which include porosity, permeability and different types of rock, sand or clay at different depths. Water's highest point in an aquifer represents the water table which may be close to the surface or deep underground, Figure 2.2.

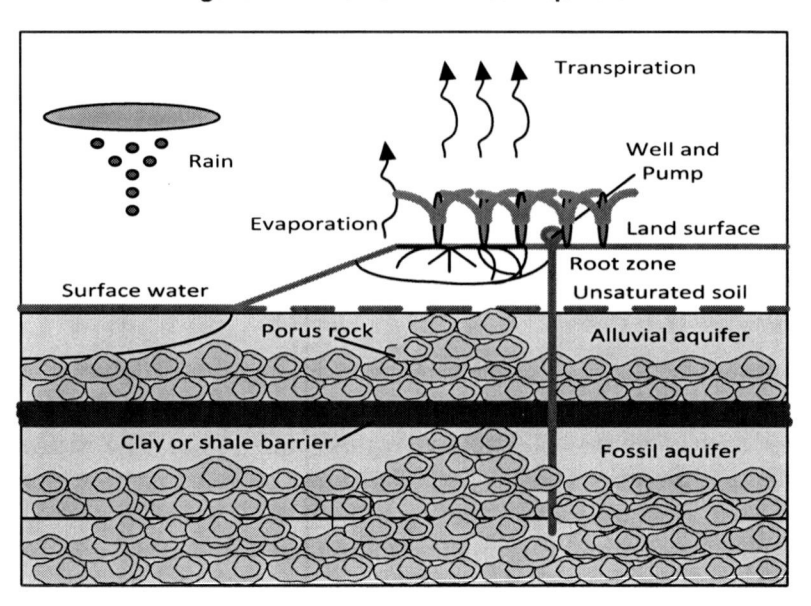

**Figure 2.2. Alluvial and Fossil Aquifers**

Aquifers close to the surface, called alluvial aquifers, may be partially recharged by annual rains. However, many alluvial aquifers refill with only 10 – 30% of annual rains as the rest runs off before filtering down to the aquifer.

Large corn producing states, California, Nebraska, Kansas, Texas, Arkansas, and Idaho, account for 53% of total U.S. irrigated acreage.[70] Much of their irrigation water depends on fossil aquifers. Fossil water was trapped thousands of years ago in ancient sediments below a layer of bedrock, shale or caliche that blocks recharge.

The large Ogallala aquifer supports the groundwater needs of eight High Plain states and runs from South Dakota to Texas. This fossil aquifer was filled with water from a glacier 25,000 years ago. Even if the Ogallala were not blocked from recharge by a layer of shale, the High Plain gets rainfall amounts less than a third of the five trillion gallons of water extracted in normal rainfall years.

Mining fossil water is identical to oil extraction – when the pool goes dry, no more groundwater is available. Mining water for irrigation or use by cities or industry causes the fossil aquifer water level to drop. Then progressively larger pipes and stronger pumps are necessary to extract water from deeper and deeper wells.

When much of the fossil water that has been in the ground for millennia is extracted, the aquifer crashes. Considerable variation occurs in aquifer death due to local geology. Some aquifers crash with 30% of the water still remaining because the water cannot be extracted due to turbidity, cave-ins, pebbles or mud. In other cases, wells must be sunk so deep, the cost of pumping the water exceeds the value of potential crops.

Many farmers on the southern end of the Ogallala aquifer in New Mexico, Texas and Oklahoma have already watched their wells go dry and their precious croplands have returned to prairie. Failing available water, rural families have to move and leave their near valueless land to nature.

The Oglala Lakota tribe in the Badlands of South Dakota lent their name to the Ogallala aquifer but they can no longer use its water because it has become contaminated. The waste streams for farming and sewage have ruined their groundwater and now they must haul water from the Missouri River 200 miles away.[71]

Cities on the High Plains will soon be forced to decide if they can afford to pipe in water or if they must disburse their residents. Tough choices are being imposed on communities today in several parts of the world including China, India and the Mid-East. Cities in the U.S. such as Aurora Colorado, Atlanta Georgia, Orme Tennessee and Palm Springs California realize now they face similar tough choices.

Globally, water tables are falling each year as fossil aquifers are being drained in key food-producing regions; especially the North China Plain, India's Punjab, Pakistan and northern Mexico. Many of these aquifers are heavily over drafted and are on a path to crash within the current generation.[72]

When irrigation lowers the water table, lakes drain and springs go dry, which may extinguish rivers at their source. Rivers simply run dry, as did the Snake River in Idaho, the Yaqui River in northern Mexico, the

Rio Grande in Texas and as San Pedro River did in Arizona in 2007.[73] Many of the people whose livelihoods depended on their river are forced to move because there may be no water for sustenance.

Other great rivers such as the southern end of the Colorado and Sacramento carry so much dissolved salt from agricultural run-off that their water is not usable for crop irrigation or human use. New genetically modified seeds for corn, soybeans and wheat are more productive but consume more water.

**Water consumption for ethanol.** An acre foot of water, 326,000 gallons, covers an acre one foot high. The USGA *Water Use Report* reported that several arid Western states such as Montana, Idaho and Arizona applied an **average of over five acre-feet** to irrigate crops while the High Plains averaged about two acre-feet of water.[74] The High Plains get about a third of their water in rain – in wet years.

A single acre of irrigated corn consumes three acre-feet – about one million gallons of water.[75] The water consumption of irrigated corn was confirmed by interviews and correspondence with both corn farmers and water companies and utilities. This research revealed that farmers often receive over three acre-feet to produce irrigated corn which is consistent with the USGA Water Use Report.

An acre of irrigated corn produces about 140 bushels of corn which yields 350 gallons of ethanol. Therefore, production of:

> **1 gallon of ethanol consumes 3,000 gallons of water.**

Each gallon of ethanol made using irrigated corn wastes 12 tons of consumptive use water. It seems inconceivable that the EPA, DOE, USGS and USDA would encourage the U.S. Congress to support an energy policy that sacrifices trillions of gallons of U.S. groundwater for a weak fuel additive.

Another way to validate ethanol's water cost uses the USDA ethanol yield per ton of corn, 89 gallons, and the approximation that producing one ton of grain requires 1,000 tons water. This computation also ignores the water cost of the rotation crop; the farm family and refinery water, and yields 11.2 tons of water per gallon of ethanol.

The huge water consumption would be trivial if, like Brazil, most ethanol feedstock land were rain fed. The National Corn Growers Association uses a U.S. Geological Survey report that estimates only 10% of U.S. corn land receives irrigation.[76] The NCGA is the same organization that argues in their self interest that burning 100 million tons of food to produce ethanol has no impact on food prices.

The USDA uses a figure of 16% of corn land as irrigated.[77] A visual inspection of ethanol refineries locations in Figure 2.3 suggests the industry and government reports for irrigated corn land may be 100 – 200% low. Their apparent reporting error creates serious consequences – Americans and politicians are unaware of the velocity at which ethanol is consuming vital non-renewable groundwater.

The USDA relies on the veracity of farmer surveys to determine irrigated land.[78] The USDA also relies on farm surveys to determine farm size. By law, large farms should not receive farm subsidies but large farms do receive millions of dollars in subsidies. Some farmers cheat and report several smaller farms. The USDA accepts survey responses because it has no authority to challenge abusers.[79]

The 100th Meridian runs through the middle of North Dakota and represents a dividing line for irrigation. Croplands west of the Meridian as well as some to the east, need irrigation regularly. The ethanol business model places ethanol refineries where most their corn feedstock comes from a radius of about 50 miles. Consequently, each refinery west of the 100th Meridian uses irrigated corn. States on the High Plains and West have recently built over 30 ethanol refineries that depend primarily on irrigated corn for feedstock.[80]

A typical 50 million gallon a year ethanol refinery consumes 150 billion gallons of consumptive use irrigation water for its feedstock. Nearly all the Western states depend on irrigation for growing crops and have ethanol refineries. California has seven refineries all using irrigation water.[81] For corn producers, not only is the corn subsidized but many farmers receive huge additional subsidies for irrigation water and the power to pump the water. These subsidies lead to tremendous water waste.

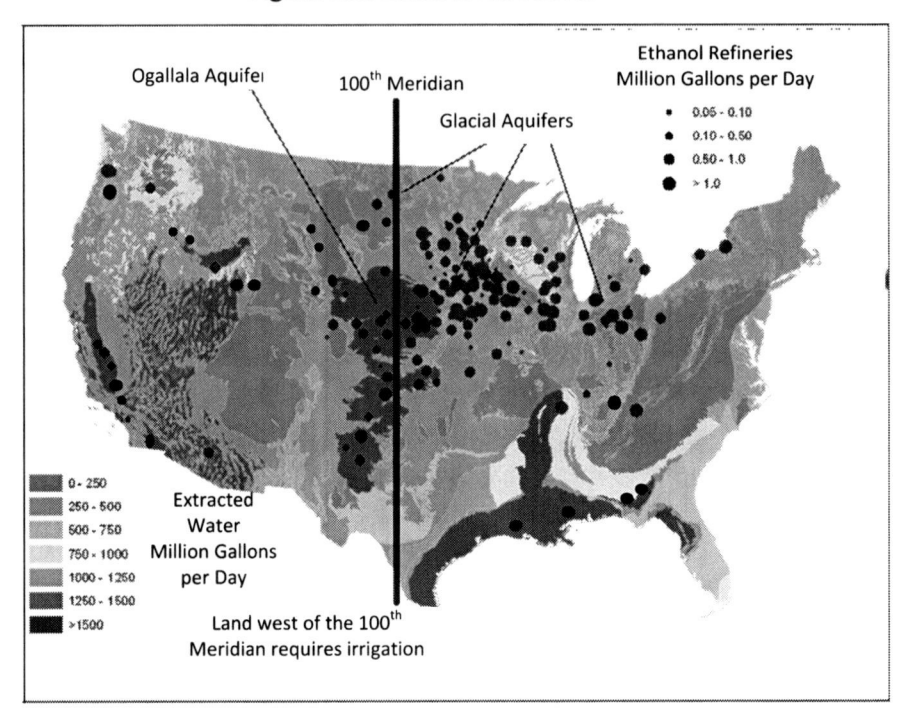

**Figure 2.3. Ethanol Refineries**

Over half of irrigation water is drawn from aquifers, many of which are nonrenewable fossil water including the Ogallala. Irrigation water for corn, similar to most food crops, cannot be too salty or polluted with industrial wastes. Wasting trillions of gallons of water on producing corn for ethanol will drain America's bread basket dry. The senseless U.S. energy policy accelerates aquifer depletion and the desertification of the High Plains and West.

### Green solar water solutions

Unfortunately neither algae nor any other technology can manufacture water. Desalinization provides at best a stopgap measure for cities desperate for drinking water. Desalinization is an order of magnitude, about 10 times too expensive in both costs and energy for providing water for irrigation.

Algae offer two strong strategies for supporting sustainable water: displacement and replacement.

**Displacement** occurs by growing foods and biofuels that would otherwise require clean irrigation water.

- **Water savings** – algae grown in closed loop algaculture systems has minimal evaporation or water loss. The water used for growth can be recycled. Replacing corn ethanol production with green solar would save 2 trillion gallons of fresh water annually.
- **Brine, waste or salt water** – algae grows well in water that would kill land crops such as brine, waste or salt water. Algae have no roots so salt ions do not interfere with water uptake or transport.

**Replacement** occurs when algae replaces polluted water with water that is sufficiently clean that it can be used for irrigation or with modest treatment, recycled for human use.

- **Clean human wastewater** – algae assimilate nutrients from water treatment plants and return the water to a near-drinkable state.
- **Clean industrial wastewater** – algal bioremediation absorb heavy metals, toxic compounds and poisons from industrial wastes. These elements and compounds can be recovered from the algal biomass and may be sold or reused. Heavy metals are easier to separate from solids, algae, than when they are in solution.
- **Clean agricultural wastewater** – conventional technologies for removing nitrogen and agricultural waste stream chemicals are very expensive and use considerable energy. Algae provide and inexpensive solution and require light energy that can be supplied with green fuels.

- **Remove pharmaceuticals from lakes** – discarded prescription and over-the-counter drugs have polluted lakes and well-water which can be cleaned with specific species of algae.
- **Industrial heated water** – many industrial manufacturing plants as well as coal-fired power plants and nuclear energy facilities use extensive fresh water for cooling. The industrial heat can be flued through algal ponds that absorb the heat and turn the heat which acts as a catalyst to speed green biomass production.

Algae also offer water and air pollutions solutions in that algae may be used to test, diagnose and monitor for a wide range of potential toxins and other contaminants. One of algae's defense strategies is to react in the presence of certain dangerous chemicals in very low dilutions. These reactions can serve as a monitoring system for water pollution.

A cultivated algal production system that recycles water may use only 0.001 as much fresh water per pound of protein production as food grains. This minimal water requirement provides significant advantage for algal production for the many global cities, towns and villages that are water insecure.

*Biofuel demand*

The world demand for liquid transportation fuels continues to escalate at about 3% annually. Increasing demand and diminishing supply drive fuel price increases. Developing countries such as China, India and Indonesia are consuming more transportation fuels which also push prices higher. Speculators are betting that future supplies will be tight and their future options will be profitable. A single production, refinery or distribution failure, storm or terrorist damage could double or triple the cost of a barrel of oil.

Both air and ocean transport will increase demand for high-value fuels over the coming decades. Fossil sources for high energy fuels are insecure and costly so biofuels that are truly renewable will provide a critical solution for liquid transportation fuel.

## Global Challenges

The U.S. energy program that pushes corn ethanol makes sense to politicians because it provides farmers with much needed income. However, the science and unintended consequences of massive corn production for non-food purposes makes no sense. If government experts and politicians were looking for a green (sustainable) biofuel feedstock, they made the worst possible choice.

The reasons algae will displace corn ethanol as the biofuel feedstock of choice is summarized in Table 2.1. Corn is not sustainable as a biofuel feedstock because it takes far too much cropland, water, fertilizers, pesticides and herbicides. Corn ethanol also costs far too much to produce in both dollars and in energy.

### Table 2.1 Corn Ethanol and Algal Biofuel Comparison

| Parameter | Corn | Algae |
|---|---|---|
| Renewable and sustainable | No – massive consumption of inputs is non-sustainable | Yes – positive ecological footprint |
| Gasoline equivalent energy | 36% less energy than gasoline | 30 to 50% more energy |
| Eflation, ethanol induced food price increases | Yes, burning food for fuel decreases supply and increases prices | No, produce both proteins and lipids |
| Decreases exports due to eflation | Yes, due to price spikes in animal feed | No, produces fodder, animal feed |
| Require $20 B in subsidies to sustain the industry | Yes, ethanol is non-sustainable without subsidies | Yes, but far lower subsidies |
| Consumes massive fossil fuels in feedstock production | Yes – tractors, trucks, harvesters and fertilizer | No – light tractors, trucks and fertilizer |

Green Algae Strategy

| | | |
|---|---|---|
| Vulnerable to a single weather event or disease vector | Yes, a drought or severe storm can destroy a crop | No, when the sun shines, algae grows |
| Need for special engine designs due to alcohol | Yes, alcohol degrades the engine | No need for special engines |
| Salt and jellylike deposits on fuel pumps | Yes | No |
| Corrosion in vehicles with fiberglass tanks | Yes, alcohol | No alcohol |
| Cannot be distributed by pipeline because it mixes too easily with water | Yes, absorbs condensation and dilutes fuel | No, can be distributed by pipeline |
| Positive energy value – yields more energy than it takes to produce | No, about energy neutral | Yes, energy positive |
| Consumes millions of acres of cropland | Yes, about 40 million acres in 2008 | No cropland |
| Consumes trillions of gallons of clean water | Yes, about 2 trillion in 2008 | No Algae clean waste water |
| Pollutes air and soils | Yes, $CO_2$, nitric oxide, fertilizers | No pollution |
| Creates dead zones in wetlands, rivers, lakes, estuaries and oceans | Yes, agricultural run-off | No, water recycled |
| Causes massive ecological destruction | Yes, extracts nonrenewable fossil water, erodes soils and pollutes wells | No ecological destruction |

| | | |
|---|---|---|
| **Systemic water pollution** | Millions of tons of fertilizers, herbicides, pesticide | **No water pollution** |
| **Erodes soils** | Yes, about 6 tons per acre per year | **No erosion** |

Corn ethanol is a renewable product in the short term but clearly non-sustainable because growing corn requires billions of gallons of fossil fuels such as gasoline, diesel and natural gas. It also consumes massive amounts of fossil water, fertilizers, herbicides and pesticides which are not sustainable.

Corn's vulnerability to production variations make it 80% likely that in the next 10 years even a modest crop disruption will lead to a catastrophic food cascade.[82] The early actions of this type of food cascade, food fights, food price escalation and food riots have occurred in 2008. Unfortunately, until the U.S. ends food destruction for fuel, food insecurity will increase.

A **food cascade** occurs when a real or perceived food shortage creates a scarcity, panic and food fights. A psychology of fear causes people to hoard food, prices spike and create an economic firestorm. Food riots and political destabilization follow. A food cascade represents a horrific but highly probable outcome from sacrificing food crops to produce biofuels.

Even if corn inputs such as land, water and fossil fuels were sustainable, the ecological damage done by massive production of corn is non-sustainable. Ethanol will drain America's groundwater dry before displacing 7% of imported oil.

America cannot afford to sustain the cost of huge subsidies while experiencing severe systemic ecological damage.

# Chapter 3. Sustainable Food – Algaculture

> Sustainable food production is no longer just a dream but an achievable objective within a decade with the first form of agriculture on Earth – nature's ingenious invention; algaculture.

Most of the world's 6.6 billion people will get their food today from foodgrains. Grains have been hybridized from grasses and take the form of wheat, corn (maize to most the world), barley, rice and beans.

Globally, the vast majority of people eat grains primarily because they are relatively cheap to grow and can be dried and stored with minimal spoilage. Dried grains are much lighter than fresh fruits and vegetables and therefore far less expensive to transport.

Meats are more expensive because meat reflects the cost of feeding animals multiple pounds of grain. Red meats such as beef and pork require about eight pounds of grain per pound of meat. This is why a vegetarian diet is much less expensive than a carnivore diet. Every meal of meat, dairy, poultry and oils represents a high multiple of the grain consumed to produce it relative to a meal of grain, fruits or vegetables.

The United Kingdom's Chief Scientific Adviser John Beddington said "It's very hard to imagine how we can see the world growing enough crops to produce renewable energy and at the same time meet the enormous demand for food." He went on to say that the world's food supply needed to double within 30 years.[83] World Bank president, Robert Zoellick also said we must double the food supply.[84]

Unfortunately, world food production may not have the capacity to increase or even sustain current production for the long term using conventional agriculture. Food production may have peaked. Economists and even ag-economists fail to factor into their business models the cost of agriculture's self-destruction, over-consumption of fossil resources and climate change. The fact that ecological costs are not objectively quantifiable does not diminish their potential impact.

*Food crop requirements*

The world's food security depends on the production of sufficient foodgrains to feed hungry populations. Growing such crops has been challenging farmers for millennia because grains require:

- **Good cropland** – high quality, fertile soil with loose dirt that allows water to percolate through it slowly and is at least 8 inches thick.
- **Climate** – temperate climate during the entire growing season from 50° to 95° F.
- **Full growing season** – 90 to 150 days of uninterrupted good growing conditions.
- **Soil moisture** – the soil must retain moisture to about 18 inches but cannot be too soggy or it will drown the plant.
- **Water delivered on time** – seeds fail to germinate or grow without sufficient soil moisture but too much rain creates mud and farmers cannot get into fields for planting, weeding, fertilizing or harvesting.
- **Absence of negative weather events** – no fierce storms, hail, frost, drought or too much rain.
- **Avoidance of pest vectors** – no invasion by rusts, fungi, molds, mildew, worms, locus, birds or bugs.

Engineering Sustainable Food

A single negative weather event or pest vector that occurs during the three to four month growing season can damage or destroy an entire crop. When heat or drought occur toward the end of the growing season as the plant begins to make its seeds, the heat or thirst interferes with fertilization causing failed seed production. Similarly, temperature spikes above or below the plant's thresholds can stunt or destroy an entire crop.

Traditional foodgrains will continue to make up a large portion of global food. Unfortunately, food grain production is not sustainable with increasing populations and climate change causing diminishing cropland, water scarcity, aggressive pest vectors and higher cost fossil fuels. Growing food grains successfully requires 15 factors. These factors in combination make foodgrains non-sustainable.

### Table 3.1 Why Foodgrains are not Sustainable

| Parameter | Explanation |
|---|---|
| 1. Diminishing cropland | Most good cropland is already in production and new cropland takes more inputs while producing less food |
| 2. Climate change | Global warming causes heat stress, drought and severe storms |
| 3. Slowing crop productivity growth | Productivity gains for critical food crops are growing at a decreasing rate. Some are diminishing |
| 4. Ag inputs insecurity | Water, fertilizers and fossil fuels are sometimes not available |
| 5. Agricultural inputs costs | Water, fertilizers and fossil fuels are becoming too expensive |
| 6. Salt invasion | Salt invasion from rising oceans that surge on croplands and dissolved salts in irrigation water are destroying large areas of cropland |

| 7. Scarce water | Overuse of irrigation drops water tables, dries up springs, reservoirs and lakes and causes aquifers to collapse |
|---|---|
| 8. Fertilizer | Overdependence on heavy fertilization costs too much and causes pollution |
| 9. Low protein biomass | Farmers must produce a large woody plant with a biomass yield of less than 10% protein |
| 10. Herbicides /pesticides | Heavy use increases costs and dependence on fossil fuels and kills honeybees |
| 11. Pesticide resistance | Hundreds of pests have developed resistance to pesticides and weeds to herbicides |
| 12. Water pollution | Agricultural runoff pollutes soils, wetlands, rivers, lakes and well-water |
| 13. Mono-cropping | Massively growing one plant variety jeopardizes the entire food system |
| 14. Ecological destruction | Growing food crops creates environmental havoc such as soil erosion, fast growing weed invasion and loss of habitat for birds, fish and amphibians |
| 15. Fossil fuels | Substantial price increases and supply insecurity threaten all agricultural inputs |

Nearly all good cropland has been under cultivation for decades because it represents a valuable resource. New land put into food crop production is less flat and less rich which makes it more demanding of inputs such as water, fertilizer, herbicides and pesticides. Unfortunately, even with these additional inputs, food productivity typically diminishes quickly on new cropland.

Global warming causes severe heat and droughts that stress, degrade and destroy food crops. Increasing ocean surface temperatures are generating more violent storms that decimate existing crops, next year's crop seeds, work and food animals and agricultural infrastructure.

Food grain productivity growth advanced rapidly during the Green Revolution by applying more irrigation, fertilizers and pesticides. The rate of increase is now slowing due to:

- Irrigation overdraft that threatens water security and drives up water and power (pumping) expenses
- Over-use of fertilizers creates plants that become less sensitive to fertilizer, drives up the costs of fossil fuels and the fertilizers itself and pollutes air, soils and water
- Pesticides and herbicides that also depend heavily on fossil fuels become inefficient due to evolving pest and weed immunity.

The unintended consequences of the Green Revolution, conventional agriculture's self destruction – cropland losses due to irrigation salt invasion, water loss, soaring fossil fuel costs, air, soil and water pollution and pest immunity – are occurring with nearly equal velocity on all continents.

Increasing costs combined with decreasing availability for key agricultural inputs diminish the potential for traditional agriculture to produce sustainable food. Farm land, labor and equipment costs are also rising rapidly which diminishes farmers' willingness to farm.

Rising oceans are currently destroying millions of cropland acres from sea salt invasion. Rising sea level projections indicate this problem more will become more severe in the coming years. Continuous irrigation causes dissolved soil salt invasion which leaves a white crust on the soil. Salt invasion from irrigation causes 25 million acres of cropland to be taken out of production every year.[85]

We can grow food without soil – hydroponics and aquaculture – but not without water. Water availability has become the most critical constraint for world food production. Over-drafting is causing groundwater tables to drop so deep that hydrology defeats farmer as the cost of pumping becomes prohibitive. Deeper and deeper wells add energy costs which eventually exceed the value of the crop.

As wells go deeper, water often becomes unfit for crop production. Dissolved salts create brine water that kills land plants. Other water contains too much boron, such as food basked of the U.S., California's Central Valley, which also destroys not just the crop but the cropland.[86]

When water tables drop, springs that start rivers, along with lakes, reservoirs and wells go dry. Aquifer overdraft, where farmers pump water at three to 50 times the recharge rate, spells doom in the near future for many aquifers.

Since much of the world's grain is produced with fossil aquifer water that is non-rechargeable, this threatens aquifer crashes within the current generation. Many fossil aquifers have already crashed which immediately stops the growing of food crops.

Glacier ice is also retreating globally, especially in Tibet, which significantly diminishes the source waters for major rivers in China, India and Pakistan.[87] In recent years, snow packs have diminished 50% in many regions such as the U.S. Rocky Mountains which causes more runoff to occur in the spring and less water to be stored for irrigation.

Rivers in all food growing continents that are critical for irrigation and transport are going dry. China's two largest rivers the Yangtze and the Yellow have both been used up and are dry before they reach the sea. Critical rivers that supply water for agriculture and cities have gone dry in India, Pakistan, the Mid-East and Africa.[88]

Much of the productivity improvement of the Green Revolution came from piling on more fertilizers. Increased fertilizer use drives up the cost of natural gas which represents 80% of the cost of fertilizer. Crops become insensitive to heavy use of synthetic fertilizers and require more applications. Increasing fertilizer costs and availability seriously constrain substantial expansion in food production.

The U.S., for example, was the leading exporter of fertilizers for decades. In 2007, American farmers imported over half their fertilizer because large increases in natural gas prices made domestic production of fertilizer uneconomic for the producers. Paying significantly higher import prices for fertilizers made producing food crops uneconomic for some farmers.

Farmers must invest all the inputs for growing food crops, especially fertilizer, to grow a plant biomass that may be less than 10% food. A corn stalk, for example, may be ten feet high and weigh 4 pounds.[89] The only food value is held in the kernels on the cob and yields only a few ounces of protein. The rest of the plant, the roots, stalk, leaves, tassel and cob consume lots of fertilizer and water to grow but represents a waste product for either food or fuel production. Algae by comparison may be 50% oil and 25% protein.

Added to the production parameters that make traditional agriculture unsustainable are a host of political and social variables. Governments may institute land reform and give valuable croplands to people with need but neither the means nor the knowledge needed for farming. Political strife or war may take all able bodied men or make farming too dangerous. Health catastrophes such as West Nile Virus, HIV/AIDS or other vector may threaten or disable farm workers.

### *Climate change and food crops*

Global warming and higher temperatures increase the need for expensive herbicides and pesticides to support current production levels. Higher temperatures enable pests to grow faster and accelerate propagation of fungi, molds, rusts and mildews. Higher temperatures often put plants in a condition of stress where they are more vulnerable to disease and pests. Spring comes earlier and fall goes on later which give pest vectors an opportunity for population explosion while they feed on food crops.

Fewer days at the cold temperature extreme means many pests who would die from freezes or frosts survive the winter and begin their attacks on plants each spring in greater numbers. The bark beetle, for example, has destroyed millions of acres of forest lands.

Farmers put tons of herbicides and pesticides on crops to control undesired weeds and pests. In 1950, there were about 10 species of insects resistant to pesticides. Today, there are over 600. Similarly, the number of weeds with herbicide resistance was near zero in 1950 but is more than 400 today.[90] Even though insecticide use has increased tenfold, crop loss from insects is now double the level it was in the 1940s – about 13%.[91]

The gnat-sized brown plant hopper is multiplying by the billions and decimating rice crops in Asia. The hopper bites the rice stem and not only damages the plant but injects a virus. The hopper can now withstand over 100 times the insecticide dose that used to kill it.[92]

Paul Ehrlich, possibly the most knowledgeable scientist in the world on the environment, notes:

> Of all human activities, agriculture arguably has the greatest environmental impact, especially the destruction of biodiversity – plants, animals and microbes that share Earth with us and upon which our lives depend. ... Agriculture itself can destabilize the very process it depends upon for success.[93]

Agricultural adds to pollution from dust, spraying and runoff that spread agricultural chemicals through wetlands and waterways. Some food crops such as corn are very inefficient users of fertilizer and absorb less than half the amount applied. Some crops may only absorb 1% of the pesticides applied. The remaining agricultural chemicals run off during storms or irrigation and poison well-water.

A 2007 Iowa Department of Natural Resources report indicates 274 Iowa waterways were seriously polluted. Fertilizer run-off causes such a problem that Iowa boasts the largest nitrate removal plant in the world.[94]

China's fertilizer run-off has turned lakes into cesspools. More than half of China's waterways are so polluted that fish are dying or water is unsafe for drinking or irrigation. More than 300 million people lack access to clean drinking water.[95] The official Xinhua news agency admits that more than one-third of the country's 85,000 reservoirs and dams have "serious" structural problems.[96]

Monocropping occurs because farmers tend to plant large areas of the same crop such as corn, soybeans, wheat or rice. Farmers plant what they think will maximize food production which leads to the same crop being grown in many regions with similar growing conditions. The entire food system is put at serious risk from a single pest vector or weather event such as the drought that ruined two thirds of Australia's expected 25 million ton wheat harvest in 2006 or the cyclone that ruined over half of Myanmar's rice harvest with 120 mph winds in 2008. Rust, mold, mildew or other pest can spread quickly due to global warming and severely damage large food growing regions where farms grow a single crop.

The intensive growing of single crops can create ecological destruction. A single corn crop is so hard on the land that it takes nature over 25 years to repair the ecological damage. Many crops, including corn, are so destructive that they must be rotated with other crops to replenish soil nutrients.

Heavy nitrogen concentrations from fertilizer runoff enable opportunistic fast-growing weeds to displace slow-growing ferns in wetlands and choke off natural water flows. Agricultural chemicals may not only destroy the habitat for animals but also severely damage internal organs and interfere with their reproduction.

Soil preparation and growing crops is fossil fuel intensive. Both the cost and availability of fuels limit future food production. Growing food requires heavy machinery powered by hungry engines that consume tons of fossil fuels. Food processing, storage and especially distribution by roads, trains, water and air are heavily reliant on fuels. A fuel supply disruption due to politics, war or natural disaster would be catastrophic for food production.

While many Americans take food security for granted, producing successful crops year after year is far from a sure thing.

### *Sustainable food production*

In order to avoid future food scarcity, food riots and the destabilization of governments, the world desperately needs more affordable food locally. A new food source is needed that not only produces the protein, carbohydrates and nutrients people need for healthy and active lives but produces the food sustainably. A nice but less critical constraint would be production of a sustainable biomass that also yields biofuel feedstock.

> Sustainable food production meets the needs of the present without compromising the ability of future generations to meet their own needs.[97]

All the variables of production must be sustainable or future generations will experience a legacy of hunger and thirst. Sustainable production depends on careful management of food production's 15 critical variables, as were described in Table 3.1 above, especially cropland, water and fossil fuels.

A sustainable food source should be robust, resource efficient and really, really simple to grow, Figure 3.1. Each of these three factors includes specific constraints.

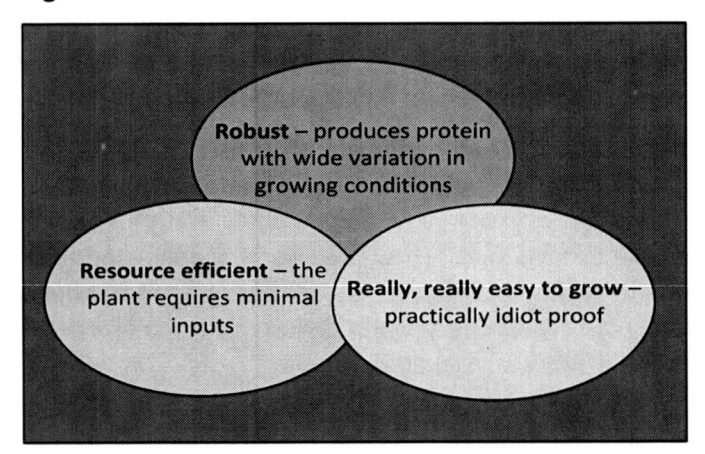

### Figure 3.1 Constraints for a Sustainable Food Source

- **Robust** – produces protein with wide variation in growing conditions
- **Resource efficient** – the plant requires minimal inputs
- **Really, really easy to grow** – practically idiot proof

**Robust** means the food source produces under a variety of conditions. An ideal sustainable food might produce proteins, oils, carbohydrates and other important nutrients effectively, all independent of the:

- **Altitude** corresponds with high, low and average temperatures.
- **Latitude** constrains the length of the growing season by the number of daylight hours.
- **Geography** means mountains, valleys, deserts or oceans.

- **Soil** associates with fertility, granularity, density, depth and ability to hold moisture.
- **Climate** is often a function of the prevailing winds and water vapor.
- **Temperature range** reflects the number of days with temperature extremes.
- **Negative weather events** such as fierce storms, hail, frost, winds, drought or too much rain.

A geographically independent plant, assuming variations in species, could be produced all over the world, even in the north latitudes, on mountains and deserts; lakes and the ocean.

Even a robust organism needs to be extremely resource efficient in order to be sustainable, especially in a global warming environment.

**Resource efficient** means the food source requires minimal inputs for effective food production. An ideal resource efficient organism would:

- Not compete with traditional food crops for scarce cropland.
- Not compete with traditional food crops or people for freshwater.
- Be space efficient and require minimal acreage for high food production.
- Not create a negative atmospheric footprint by producing carbon dioxide, nitric oxide or other greenhouse gases.
- Not create a negative soil footprint such as putting extensive fertilizers, herbicides and pesticides in the soil.
- Not create a negative water footprint by polluting surface and ground water with soil runoff and agricultural chemicals.
- Demand minimal fossil fuel energy for its food production.
- Consume minimal fertilizers, herbicides and pesticides.
- Need minimal packaging, storage and transportation.

Resource efficiency requires that a food source must grow and develop in Earth ecosystems that do not compete with traditional foodgrains or other crops. Resource efficiency also coserves precious resources such as water and the ecology for the needs of families, cities, businesses as well as traditional food crops.

The food source must be **really, really simple to grow** because millions of farmers will have to adopt new methods. Jeffry Sachs proposes that an effective R&D model for the developing world is R3D; research, development, demonstration and diffusion.[98] Basic research can make the food source available but adoption requires that the plant be very easy to demonstrate and to teach others to grow.

Professor Emeritus Everett Rogers from Stanford studied the diffusion of innovation for decades and documented the diffusion of hybrid seed corn and the incredibly slow progress in diffusing new agricultural methods. He concluded that people's attitude toward a new technology is a key element in its diffusion. Roger's Innovation Decision Process theory states that innovation diffusion is a process that occurs through five stages: knowledge, persuasion, decision, implementation and confirmation.[99]

Jeffry Sachs' R3D model aligns with Rogers' findings to indicate that simplicity will be critical for effective demonstration. A food source that grows sufficiently easy for wide diffusion and adoption must be easy to grow farmers and support persons such as family members:

- Experience a one-day learning curve – see and immediately do.
- Have to learn only a few basic actions and few new technologies.
- Are able to learn visually so language will not be an obstacle.
- Are able to invest labor and simple tools rather than buying complex, expensive technology.
- Know how to make the plant grow and produce even if they do not understand how and why it produces.

A plant that is robust, resource efficient and really, really simple to grow presents a workable model for sustainable food production.

### *Algaculture – sustainable production*

Sustainable food may be produced with nature's first plant culture on Earth; algae. This organism, the size of a nanoparticle, enables the oldest form of agriculture on Earth, algaculture. Nature's first form of agriculture transforms solar energy into high energy green biomass using photosynthesis. This green solar process generates a flexible plant biomass that may provide energy for:

- **People** – protein in food
- **Animals** – protein in fodder
- **Fowl** – protein for birds
- **Fish** – protein in fish feed
- **Plants** – nitrogen fertilizer
- **Fire** – high energy algal oil for cooking and heating
- **Cars** – high energy algal oil for transportation
- **Trucks and tractors** – high energy green diesel
- **Planes** – high energy jet fuel

Algaculture or green solar produces food, fuel, coproducts and pollution solutions sustainably without sacrificing cropland, fresh water or fossil fuels and provides a positive ecological footprint. Algaculture applies a seemingly novel but actually nature's ancient strategy for growing food; the use of water-based plants with special characteristics.

Traditional agriculture is non-sustainable due to its heavy reliance on non-sustainable natural resources. The substantial costs of agricultural inputs make food unaffordable for many, especially those already suffering from malnutrition. It also depends on relatively stable weather and requires considerable labor or heavy equipment.

Organic farming minimizes reliance on fossil fuels by avoiding synthetic fertilizers, herbicides and pesticides but still depends upon considerable fossil fuel to operate farm machinery. Organic farming is not sustainable because there is insufficient cropland to produce enough green manure, the vegetative wastes for fertilizers. In addition, organic farming often suffers from low productivity, pest and weed invasions and takes nearly as much fossil water as traditional agriculture. Organic farming is highly vulnerable to unstable weather.

Algaculture makes use of Earth's oldest natural food culture and avoids the need to use of valuable cropland, freshwater and fossil fuels. Algaculture produces both food and fuel using sunshine, waste or brine water and $CO_2$. and is largely weather insensitive. Even though algal cells are small, a single cell can produce over a million offspring in one day. Under proper growing conditions, some species multiply their biomass two, three or even four times a day. They flourish when the sun shines and rest at night and on cloudy days.

The food value of this plant has been known for centuries and the food potential for at least 100 years. Unfortunately, this organism's potential to be the sustainable food plant of choice holds true currently only on paper and in the laboratory. Three key issues need to be solved before algae can serve as a solution for sustainable food.

1. Commercial scale biomass production on large farms.
2. Small scale biomass production with low technology for homes, communities and villages.
1. Cell wall issues for easy lipid and protein extraction and digestibility for humans and animals.

Possibly the only good to come from increasing fossil fuel prices is that high prices motivate substantial investments in commercial scale algal production for biofuels. Micro-production will be straightforward once commercial scale production becomes viable. While this may sound backwards in a development sense, large-scale production receives 98% of the R&D money because the payoff is so large.

Currently algae cell walls behave more like the shell of a nut while biotechnology and chemical engineering will make them act more like a grape. Algae can be hand-pressed to extract lipids but hand pressing is inefficient and wastes lipids left in the biomass.

The challenges facing algae production are far less rigorous than man's first flight. Few were interested in these challenges when food and fuel were plentiful and cheap. Fortunately, algae's potential for low cost sustainable biofuel production has created an intense green biomass race because the potential rewards are stratospheric.

Resolving production and cell wall issues involve a host of variables discussed in Chapter 6. When production and cell wall challenges are solved in the next few years, algae will be ready to serve as a sustainable food because it is robust, resource efficient and farmers can be trained to grow it easily.

Algae are incredibly **robust** because, in closed loop algaculture systems, the plant grows independent of:

- **Altitude** – It grows everywhere on the planet including in snow on mountain tops and under the ice shelf in Antarctica.
- **Latitude** – It multiplies the biomass multiple times daily when there is sun so it has no growing season. Some algaculture systems may not produce effectively in winter unless artificial light is used. Cold locations may select cold productive species or use geothermal heat to warm greenhouse production.
- **Geography** – It may be grown on mountains, deserts, rooftops, garbage dumps, wasteland or oceans.
- **Soil** – It does not need soil for growth since it grows in waste, brine or salt water.
- **Climate** – It can be grown in a closed system independent of climate. In extreme climates, producers may need to grow one species in the fall and spring and another in the summer.
- **Temperature range** – Various species grow in most temperature ranges.
- **Negative weather events** – Closed systems are largely weather-event independent. Storms or drought have minimal impact on production because the biomass is sheltered from the weather.

Algae are **resource efficient** because they learned to survive in the brutally harsh conditions on Earth 3.5 billion years ago. The plant requires minimal inputs for successful growth. It:

- Needs no cropland and prefers deserts where food crops do not grow. Marine species may be grown in estuaries, bays and oceans.
- May be grown in water that is polluted with industrial wastes, human wastes or in brine or saline water. Algae grown for food will probably be grown with salt, brine or freshwater unless technical breakthroughs occur in separating waste products such as heavy metals from the biomass.

- Creates substantial biomass production in a very small space.
- Creates a positive atmospheric footprint by feeding on $CO_2$ and water and giving off lots of $O_2$ as it makes plant biomass.
- Creates no soil ecological footprint since it is grown in closed containers and any fertilizer is recycled with the growing water.
- Creates a neutral water footprint since the water is recycled in closed containers. Some evaporation loss occurs in warm climates.
- Demands minimal fossil fuel energy for biomass production. Renewable sources such as algal oil, solar or wind may provide sufficient energy for production.
- Consumes minimal fertilizers, herbicides and pesticides. No inputs are wasted since nutrients can be recycled with the growing water.
- Needs minimal transportation since it can be grown locally.

Annual protein production per acre for soybean is 356 pounds, rice 265 pounds and corn 211 pounds. Annual protein production per acre of algae is estimated to be about 30,000 pounds per acre.[100] In addition, algae provide a strong set of vitamins and minerals not found in land plants.

Even though algae may be robust and resource efficient, adoption by farmers around the world depends on demonstration and diffusion. Models developed by E. F. Schumacher, *Small is Beautiful*, Jeffrey Sachs, *Common Wealth: Economics for a Crowded Planet*, UNICEF, The Hunger Project and The Heifer Project provide templates for demonstration and diffusion.

Farmers may have to learn how to balance water acidity, the equivalent to testing a swimming pool which is designed to kill algae with acid. Algal are robust in many ways but are pH sensitive.

Training materials will need to be available visually such as a simple picture book. Naturally, farmers will be able to ask questions about their concerns. Fortunately, farmers will be able to produce foods and possibly cooking fuel, without needing to know exactly why the process works.

Engineering Sustainable Food

Farmers must be able to easily learn to grow the biomass and extract the oils for cooking stoves and the protein for food or fodder. Micro-algaculture systems might be demonstrated in a few hours. Farmers could learn to produce biomass in one day. With a smart rollout strategy, farmers can take a quick look at a demonstration unit and immediately begin producing algae. They may need help building a local production system and coaching through the first several harvests.

*Green Revolution – three waves*

Growing sustainable food, the heart of the Green Algae Strategy, represents the third wave in the Green Revolution.

The first Green Revolution wave saved humanity, at least temporarily, from the Malthusian trap where population growth was predicted to exceed food supply. Plentiful cheap food sparked huge population growth. Countries and their citizens became dependent on their expectation of ever increasing agricultural productivity, Table .3.1.

### Table 3.1 Three Waves of the Green Revolution

| Period | Agricultural Innovations | Consequences |
|---|---|---|
| Green Revolution Wave 1<br>1950 – 1970s | **Widespread adoption:**<br>• Fertilizers<br>• Irrigation<br>• High yield seeds – hybrids | • More food<br>• Explosive population growth<br>• Dependence on increasing agricultural productivity |
| Green Revolution Wave 2<br>1980 – 2000s | **Widespread:**<br>• Heavy fertilization<br>• More pesticides and herbicides<br>• Severe irrigation water over-drafting | • Severe water pollution<br>• Water tables plummet<br>• Aquifers crash<br>• GMO seeds needed more water |

## Green Algae Strategy

| | | |
|---|---|---|
| | • More GMOs (genetically modified organisms)<br>**Failure to:**<br>• Curb population<br>• Curb meat consumption<br>• Invest in agricultural research | • Pests and weeds became resistant<br>• Meat consumption per person doubles<br>• More dependence on increasing agricultural productivity<br>• Ag productivity decreased in many areas |
| **Green Revolution Wave 3**<br><br>**Green Independence**<br><br>**2000 – 2020s** | **Widespread:**<br>• Expansion of Rev 2<br>• Recognition of peak agriculture<br>• R&D on non-agricultural foods<br><br>**Keen awareness of:**<br>• Algaculture, green solar – nature's first culture<br>• Carbon, water and ecological footprint<br>• Pollution threats to human health, economics and vitality | • More of Wave 2<br>• Climate change impacts<br>• Ag productivity down<br>• Crop failures from heat<br>• Aquifer crashes<br>• Widespread food insecurity<br>• Food price increases<br>• Increased food imports for China, India, etc.<br>• Food riots<br>• Waste and pollution<br><br>**Declines in:**<br>• Grain production<br>• Water over-drafting<br>• Meat consumption per person<br>• Waste and pollution |

The second wave continued agricultural innovation and productivity increases. However, gains in food production were artificially propped up by ever-increasing fertilization, extensive use of pesticides and herbicides and severe over drafting of non-renewable groundwater. The second wave struck a productivity ceiling due to climate change –

more heat, drought, evaporation, pests, severe storms and wildfires combined with less irrigation water.

The third wave added amplitude to the impacts of global warming and water scarcity. The third wave experienced substantial increases in energy costs, pollution and the use of food for fuel. In addition, food riots, national disasters, aquifer crashes and political instability made food production impossible in some growing regions.

In response, the third wave Green Revolution employs novel tactics:

1. Introduction of green solar, algaculture; nature's first culture on Earth.
2. Green Independence with a plan to end oil imports, end need to use fossil fuels, recapture fossil carbon and end hunger.
3. Consumer education in food and ecological sustainability designed to help consumers make ecologically smart consumption choices.
4. Wide-spread adoption of algaculture for sustainable food, fuels, pollution solutions and other products.

The first two waves focused almost entirely on maximizing agricultural production while ignoring the true costs of over consuming non-renewable resources. The third wave continues to expand food production but operates to reduce consumer demand for foods that have high costs in non-sustainable resources such as water, fossil fuels, fertilizers and pollution.

Engineering sustainable food presents a rigorous challenge. Finding solutions to the issues of commercial and small-scale production combined with resolving cell wall issues will position algae for wide scale adoption.

The next three chapters provide an introduction to algae and green biomass production. For those who may be eager to see algal products and solutions, you may jump to Chapter 7.

# Chapter 4. Why Algae?

The first algal (singular) cell was among the earliest life-forms on Earth, probably about 3.5 billion years ago in oceanic environment synthesized by abiotic, high energy processes including lightning, ultraviolet radiation and pressure shock. The atmosphere was anaerobic with high levels of methane, hydrogen and ammonia but no oxygen.

Algae break the rules for plant classification because they evolved in many different forms – cells, multicellular plants, bacteria and in nearly infinite combinations. While the various species share certain characteristics, different algae display extraordinary variety in shape, size, structure, composition and color.

Algae are differentiated from other plants because they generally:

- Display the ability to perform photosynthesis with the production of molecular oxygen, which is associated with the presence of chlorophyll *a, b* or *c;*
- Do not have specialized transport tissues or organs consisting of interconnected cells that move nutrients and metabolites among different sites within the organism;
- Reproduce sexually or asexually to produce gametes that generally are not surrounded by protective multicellular parental tissue.[101]

Land plants evolved from algae about 400 million years ago.[102] Land plants have specialized cells for moving nutrients and for reproduction that algae do not need. Algae are distinguished from the higher plants by a lack of true roots, stems or leaves. Some seaweed appears to have leaves or trunk but they are pseudo leaves made up of the same cellular structure as the rest of the plant.

Many species are single-celled and microscopic including phytoplankton and other microalgae while others are multicellular and may grow large such as kelp and Sargassum. Phycology, the study of algae, includes the study of prokaryotic forms known as blue-green algae or cyanobacteria. Some algae also live in symbiosis with lichens, corals and sponges. The basic single-celled organism, algae, has the general appearance illustrated in Figure 4.1. The University of Montreal, U.C. Berkeley, University of Texas and others host culture collections of algal species with descriptive details and pictures.[103]

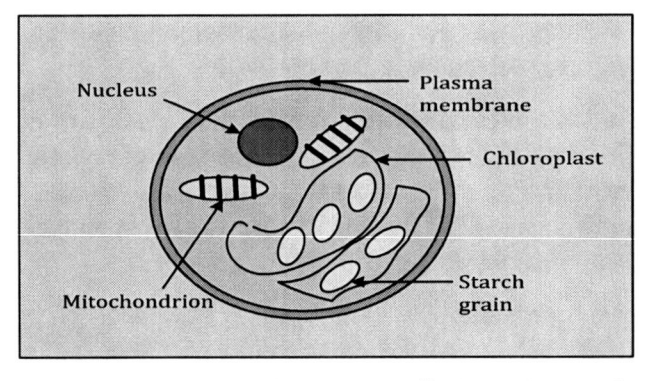

### Figure 4.1 Algae Cell

Eukaryotic green algae (Greek for "true nut") plants have cells with their genetic material organized in organelles. They create discrete structures with specific functions and have a double membrane-bound nucleus or nuclei. The prokaryotic cells of blue-green algae, cyanobacteria, contain no nucleus or other membrane-bound organelles.[104]

The major groups of algae have been distinguished traditionally on the basis of pigmentation, shape, structure, cell wall composition, flagellar characteristics and products stored. Algae display so many variations, even within each species, that they express exceptions to nearly every classification rule.

Algae can be lively little critters even though they are not animals. Many can swim such as dinoflagellates that have little whip-like structures called flagella which pull or push them through the water. Some algae squish part of their body forwards and crawl along solid surfaces.

Other species are made of fine filaments with cells joined from end to end. Some clump together to form colonies while others float independently. Seaweeds may grow in nearly any shape such as cones, tubes, filaments, circles or may imitate the shape of land plants. Seaweeds developed in parallel evolution with land plants.

## Algal Cell Walls

Major steps in cell complexity occurred with the evolutionary progression from a virus to bacterium and then from the prokaryotic cells of bacteria to the eukaryotic cells of algae. Cell walls enable algae to protect itself from the surrounding environment, typically water and pressure, called osmotic pressure.

Cell walls regulate osmotic pressure produced by water trying to flow in or out of the cell through its semi-permeable membranes due to a differential in the solution concentrations. Algae typically possess cell walls constructed of cellulose, glycoproteins and polysaccharides while some species have a cell wall composed of silicic (silicon) or alginic acid.

Red algae, for example, are a large group of about 10,000 species of mostly multicellular, marine algae, including seaweed. These include coralline algae which live symbiotically with corals, secrete calcium carbonate and play a major role in building coral reefs. Red algae such as dulse *(Palmaria palmata)* and laver (nori/gim) are a traditional part of European and Asian cuisine and are used to make other products such as agar, carrageenans and other food additives.

The broad algae classification includes:

- Bacillariophyta – diatoms
- Charophyta – stoneworts
- Chlorophyta – green algae
- Chrysophyta – golden algae
- Cyanobacteria – blue-green
- Dinophyta – dinoflagellates
- Phaeophyta – brown algae
- Rhodophyta – red algae

**Diatoms, stoneworts and dinoflagellates**

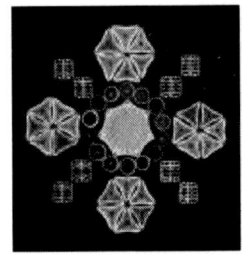

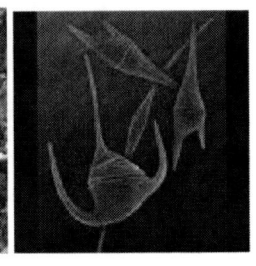

*Oxygen and nitrogen*

Fossils several billion years old suggest blue-green algae, Cyanobacteria, used their photosynthesis ability to convert the early Earth atmosphere which was devoid of $O_2$ to the presence of $O_2$. Algae converted the Earth's atmosphere to oxygen one tiny cell at a time – a process that took over a billion years. Available $O_2$ dramatically changed life on Earth and created an explosion of biodiversity which led to the evolution of land plants, animals and humans.[105]

Green algae evolved with chloroplasts which enables photosynthesis and greatly enhances available $O_2$. Blue-green algae have received most of the recent research because many scientists trained in bacteria research have begun studying the commercial value of the plant classified as both a blue-green algae and bacteria; cyanobacteria.

Prochlorococcus, a blue-green algae may be the smallest organism on Earth, only 0.6 microns (millionths of a meter), but is one of the most abundant organisms on the planet. A single drop of water may contain more than 100,000 of these single-celled organisms. Trillions of these minute cells make up invisible forests and provide about half the photosynthesis in the oceans.[106] Even though all algae species combined represent only 0.5% of total global biomass by weight, algae produce about 66% of the net global production of oxygen on Earth – more than all the forests and fields.[107]

Algae's ability to sequester $CO_2$ and produce massive amounts of $O_2$ has prompted some scientists to theorize that propagating algae in large ocean dead zones may be a way of sequestering $CO_2$ and adding to atmospheric oxygen. The ocean iron fertilization, OIF, process seeds iron in open oceans to feed phytoplankton that grow quickly and consume millions of tons of $CO_2$. The plankton bloom, mature and die and then sink to the ocean, carrying carbon with them. Ken Buesseler, a scientist of marine chemistry and geochemistry at Woods Hole Oceanographic Institution in Massachusetts along with other scientists are trying to get approvals and funding for more research.

Critics worry that seeding the ocean with large volumes of iron might have unintended consequences. In a special report, the Intergovernmental Panel on Climate Change called ocean iron fertilization "speculative and unproven and with the risk of unknown side effects."[108] However, several companies are making plans to implement ocean carbon sequestration projects, including Climos and Australia's Ocean Nourishment. Additional research is clearly needed.

Algae use nitrogen to manufacture amino acids, nucleic acids, chlorophyll and other nitrogen compounds. Cyanobacteria are able to fix nitrogen absorbed from the air, as well as from water, in a process known as diazotrophy. Since the atmosphere is 70% nitrogen, nitrogen fixing is a strong competitive advantage for growth because water-based nitrogen is often limited.

Nitrogen fixing also means that the plant biomass has value as a low energy input, high nitrogen fertilizer because algae fixes nitrogen naturally, without added energy. About 80% of the cost of commercial synthetic fertilizers comes from the energy, typically natural gas, used to extract nitrogen from the air.

Algae, often called microscopic phytoplankton, grow in most bodies of water, moist places, on and in trees and even in rocks. This little plant provides the foundation for the marine food chain feeding both microbial and animal plankton; zooplankton. Subtract algae and phytoplankton from the water column and fish, shellfish, reptiles and other aquatic creatures cannot survive.

## *Variation*

Algae range from microscopic single-celled organisms to multicelled organisms and to 180 foot kelps. These plants may be found all over the world in marine and fresh water environments – nearly any moist environment. Terrestrial algae may also be found dried in all types of soils where they can capture nitrogen from the air that can be used through the roots of plants. They may be free-living or live in symbiotic association with a variety of other organisms such as lichens and corals.

**Algal Shapes**

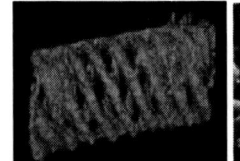

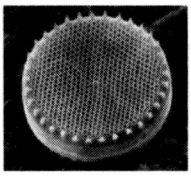

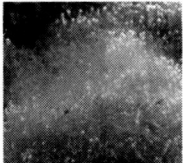

Each species may exhibit multiple strains with unique characteristics. A single strain may display completely different structural expression and composition in different growing conditions with variations in light, temperature, nutrients, mixing or water pH.[109]

## *Aquatic Species Program*

DOE sponsored an extraordinary project for 17 years called the Aquatic Species Program at the National Renewable Energy Laboratory (NREL) in Golden, Colorado. Scientists identified thousands of species that offered extraordinary potential for food, renewable fuels and many other applications such as cleaning polluted water.[110] Their interest came from the nature of microalgae:

## Why Algae?

Microalgae are remarkable and efficient biological factories capable of taking a waste (zero-energy) form of carbon ($CO_2$) and converting it into a high-density liquid form of energy (natural oil).

High energy oil-producing algae can be used to produce biodiesel, a natural oil that is emerging as a viable option for diesel engines. Algal biodiesel makes available high-volume re-use of $CO_2$ generated in power plants. It is a technology that marries the potential need for carbon sequestration in the electric utility industry with the need for clean-burning alternatives to petroleum in the transportation sector.[111]

A Look Back at the U.S. Department of Energy's Aquatic Species Program: Biodiesel from Algae

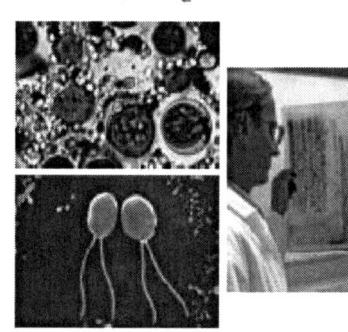

*Close-Out Report*

However, the *Close out Report* summarized years of failure to grow algae due to low yields, unstable algae cultures, harvesting difficulties, pond design and impractical photobioreactors (closed growing containers).

The program's original goals of genetically manipulating algae so that they produce more lipids were not successful. The researchers failed to identify the magic "lipid trigger" they were looking for. The report concluded that:

> Although much remains to be done, significant progress was made in the understanding of environmental and genetic factors that affect lipid accumulation in microalgae, and in the ability to manipulate these factors to produce strains with desired traits. The evidence for a specific lipid trigger is not overwhelming.[112]

A related finding was equally distressing.

> One of the most important findings from the studies on lipid accumulation in the microalgae is that, although nutrient stress causes lipid to increase in many strains as a percentage of the total biomass, this increase is generally accompanied by a decrease in total cell and lipid productivity.[113]

The path forward based on the Aquatic Species Program and the experience of other research in algal production shows that robust algal species for biofuel production need the following properties:

- High and constant lipid content
- Grow continuously (requires overcoming the stability problem common to algae cultures)
- High photosynthetic efficiency resulting in high and constant biomass productivity
- Capable of withstanding seasonal climatic differences and daily changes in temperatures
- Easy to harvest and to extract lipids (soft or flexible cell walls)

Regrettably, the political winds that blew in with the EPA Clean Air Act blew out all renewable biofuels that competed with corn in 1995. The *Close Out Report* shows excellent progress in identifying plant characteristics. Scientists working on renewable algal biomass for energy production were reassigned. Similarly, grant monies for algal biomass research at universities evaporated so researchers working on algae had to make a choice of working for free or conducting research on the only politically acceptable biofuel plant; corn.

### *$CO_2$ conversion*

Algae survived over 3 billion years on Earth because it learned to grow biomass quickly in a wide variety of conditions. Plants use the sun's energy through photosynthesis to convert sunlight into chemical energy, Figure 4.2. They convert inorganic substances such as carbon, nitrogen, phosphorus, sulfur, iron and some trace elements into organic matter such as green, blue-green, red, brown or other color biomass.[114]

Why Algae?

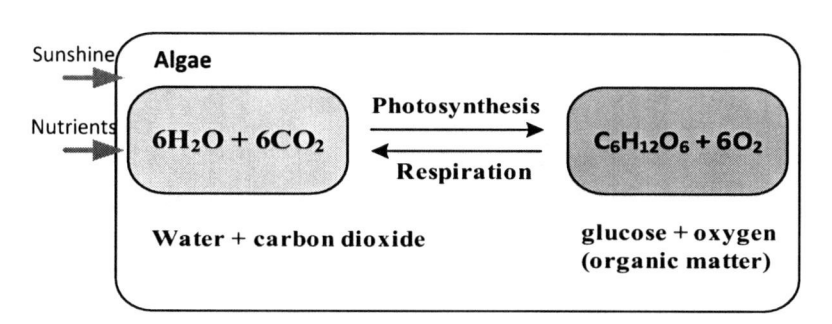

### Figure 4.2. Algae Converts Carbon Dioxide to Oxygen

```
Sunshine   Algae
                        Photosynthesis
Nutrients  6H₂O + 6CO₂  ──────────────→  C₆H₁₂O₆ + 6O₂
                        ←──────────────
                          Respiration

           Water + carbon dioxide        glucose + oxygen
                                         (organic matter)
```

Many algae are able to synthesize both organic and inorganic substances and some can synthesize without sunlight.[115] Several species of the freshwater or soil diatom Navacula are able to grow with glucose as the sole carbon source in light or in the dark.[116]

Algae feed on the greenhouse gas $CO_2$ and convert it to simple plant sugars and lots of $O_2$. Water stores little dissolved $CO_2$ naturally so cultivated algae need added $CO_2$ for food. Photosynthesis takes in $CO_2$, nutrients and water and produces the algal biomass with proteins, carbohydrates and lipids. The process releases considerable oxygen to the atmosphere.

*Colors*

The green often associated with algae comes from chlorophyll but algae also contain pigments of many colors, especially cyan, red, orange, yellow, blue and brown. Some varieties are colorless. Green algae appears green because green is the only color of light it does not absorb. Red algae absorb a full spectrum of colors and reflect red.

Algae use pigments to capture sunlight for photosynthesis but each pigment reacts with only a narrow range of the spectrum. Therefore, algae produce a variety of pigments of different colors to capture more of the sun's energy. Algae channels light into chlorophyll *a*, which converts light energy into high energy bonds of organic molecules.[117]

**Green, Blue and Red Algae**

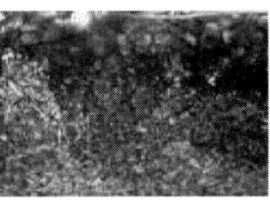

Algae provide color to herbivores that feast on them. Algae give the greenish cast to the white fur of the well-known giant sloths. Algae live in the hollow hairs of polar bears and provide the pink pigment for flamingos, which they consume in both shrimp and algae.[118] Similar carotenoids give the pink pigmentation to salmon.

Arizona's Palo Verde nuclear power plant attracted a pink flamingo to its cooling ponds several years ago. The poor bird turned white and created worldwide press speculation about possible radiation leaks. Fortunately, a biologist figured out the ponds lacked sufficient beta-carotene in the algae to sustain the bird's pink coloration. The flamingo flew to another pond with algae and quickly regained its pinkness.

Algae may grow in symbiosis with fungus to create lichen – the colorful rough material on the sunny side of rocks and trees. Algae and the fungus share a mutual dependence as the algae produces food for both plants and in exchange, gets water and minerals from the fungus. The fungus also provides critical protection against desiccation – drying and dying in the sun.

The use of algae-lichen plants for pigments and dyes pre-dates Julius Caesar. The classic red color of Roman tunics came from pigments extracted from lichens known as urchilles. Roman women valued the plant and used it as rouge to give their faces more color.

*Growth*

In nature, algae's greatest strength acts as a weakness. Fast growth shades new and prior plants from sun light. The underlying plants are shaded or receive too little light for photosynthesis and die.

Why Algae?

Algae evolved their rapid growth as a defense strategy. Since herbivores could decimate their population, algae developed the ability to combine small cell size with high velocity cell division to generate large commune populations. Predators could not find and eat all the algal cells assuring viability of the commune. Other species developed the ability to grow in cold seasons when herbivore predators were relatively low.

Another unusual strength works against algae in natural habitats. The high protein composition, often around 50% of the biomass, means the plant begins breaking down faster than shrimp – which for practical purposes means immediately. Cultivated algae harvest occurs daily but algae in natural settings attract bacteria that break down the algal biomass and give off hydrogen sulfide and ammonia, two off-putting fragrances.

**Red, Green and Yellow Lichen**

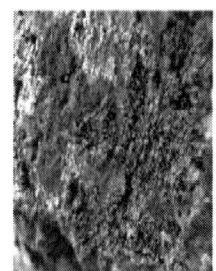

Consequently, people tend to think of algae based on its natural settings where it often presents itself as smelly green slime. In contrast, cultivated algae give off rich $O_2$ which smells similar to walking through a redwood forest (without the trees).

Algae are infamous for causing problems in public waterways and in personal pools, ponds, pots and aquariums. Algae's tolerance for a wide range of growing conditions means it demonstrates its resilience and fast growth in any moist or wet area that gets sunlight. As a result, algal research has focused heavily on trying to kill, control or remove the productive green biomass versus cultivation.[119] As a consequence, survey research indicates over 95% of people view algae as a pest.[120]

## *Structure*

Algae range from single-cells to large multicellular plants in millions of shapes, colors and sizes. They store energy as starch within cell walls composed of an outer pectin layer and an inner cellulose layer.

Most algae have microfibrillar cell walls made of cellulose that only ruminants (animals with four stage stomachs) such as cows, sheep and goats can break down. Strong cell walls evolved in algae as a brilliant defense mechanism that allows other organisms to eat the plant but not be able to digest it – break down the cell walls and extract nutrients. Algae simply pass through animal stomachs, feasting on nutrients along the way and then are secreted in a food rich environment – manure.

The cell wall defense strategy also served to distribute algae far and wide. While most algal species have hard cell walls but some algae, such as the cyanobacteria and health food Spirulina, have soft cell walls that are partially digestible by humans.

Biotechnology is making substantial progress in growing strains of algae with softer cell walls. Soft cell walls are critical for low-cost extraction of lipids as well as to facilitate human consumption and digestion.

Marine macro-algae, often called seaweeds, are biologically similar to fresh water algae. Salt water algaculture compares favorably with freshwater in terms of growth speed, robustness and biomass composition. More research has occurred with freshwater algae probably because freshwater environments are easier to maintain.

**Marine Algae – Yellow, Green and Red**

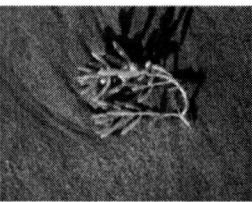

Some scientists believe that algal production for both biofuels and food may imitate fish farming in estuaries, bays and the ocean. The

Why Algae?

advantage to a cultivation process similar to aquaculture is the substantial reduction in cost for containers.

## *Composition*

Algal biomass composition includes primarily lipids, used to produce biofuel, and starches and proteins with food value.

- **Lipids** are long carbon chain molecules. Lipids store energy for the plant and serve as the structural components of cell membranes.
- **Proteins** are large organic compounds made of amino acids arranged in a linear chain connected by peptide bonds. The plant's genetic code determines the sequence of the amino acids but nutrient limitations may cause changes to the production of amino acids. Most proteins are enzymes that catalyze biochemical reactions and plant metabolism. Other proteins maintain cell shape and provide signaling functions.
- **Starches** are complex carbohydrates which are insoluble in water. Plants use starches to store glucose, plant sugar.

The composition variation among species varies tremendously. Some algae hold 80% lipids while others are 60% protein and still others are 92% carbohydrates, Figure 4.3. Species selection is critical not just for the desired composition but for a host of composition and grow variables that vary widely across species and strains.

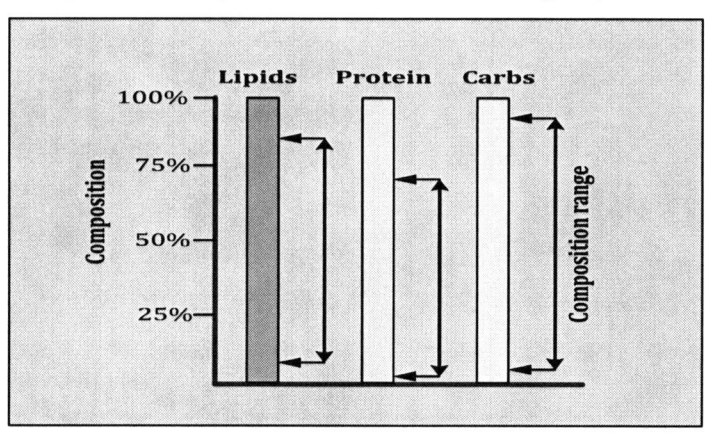

**Figure 4.3 Composition Variation across Algal species**

When algae are nutrient limited, such as nitrogen, phosphorous or sulfur, they decrease the amounts of essential polyunsaturated fatty acids produced and may yield lower quality protein with fewer amino acids. Nutrient deprivation may cause algae to increase lipid production but unfortunately, nutrient deprivation often slows or halts propagation and growth. Bioengineers are working on algae that increase lipids without nutrient deprivation.

Algal varieties offer an almost limitless combination of features. Special attributes are being enhanced through selection screens for naturally occurring organisms, bioengineering and hybridization.[121] Each species has a different proportion of lipids, starches and proteins, Table 4.1. Some algae are high protein and others are mostly starches or lipids.[122]

Table 4.1 Composition of Various Algae (% of dry matter)

| Algae | Lipids | Protein | Carbohydrates |
|---|---|---|---|
| *Anabaena cylindrica* | 4–7 | 43–56 | 25–30 |
| *Aphanizomenon flos-aqua* | 3 | 62 | 23 |
| *Arthrospira maxima* | 6–7 | 60–71 | 13–16 |
| *Botryococcus braunii* | 86 | 4 | 20 |
| *Chlamydomonas rheinhar.* | 21 | 48 | 17 |
| *Chlorella ellipsoidea* | 84 | 5 | 16 |
| *Chlorella pyrenoidosa* | 2 | 57 | 26 |
| *Chlorella vulgaris* | 14–22 | 51–58 | 12–17 |
| *Dunaliella salina* | 6 | 57 | 32 |
| *Euglena gracilis* | 14–20 | 39–61 | 14–18 |
| *Prymnesium parvum* | 22–38 | 30-45 | 25-33 |
| *Porphyridium cruentum* | 9–14 | 28–39 | 40–57 |
| *Scenedesmus obliquus* | 12–14 | 50–56 | 10–17 |
| *Spirulina platensis* | 4-6 | 46-630 | 8-14 |
| *Spirulina maxima* | 6-7 | 60-71 | 13-16 |

Why Algae?

| | | | |
|---|---|---|---|
| *Spirogyra* sp. | 11–21 | 6–20 | 33–64 |
| *Spirulina platensis* | 4–9 | 46–63 | 8–14 |
| *Synechococcus* sp. | 11 | 63 | 15 |

Algal-oils are extremely high in unsaturated fatty acids and various algal-species provide:

- **Linoleic acid**, an unsaturated omega-6 fatty acid and finds uses for soaps, emulsifiers, quick-drying oils and a wide variety of beauty aids. The moisture retention properties are valued skin remedies used for smoothing and moisturizing, as an anti-inflammatory and for acne reduction.
- **Arachidonic acid**, an omega-6 fatty acid also found in peanut oil. This product moderates inflammation and plays an important role in the operation of the central nervous system.
- **Eicospentaenoic acid,** an omega-3 fatty acid and gives the same benefits as fish oil. Research suggests that EPA may affect depression and moderate suicidal behavior.[123]
- **Docasahexaenoic acid**, an omega-3 fatty acid generally found in fish oil and is the most abundant fatty acid found in the brain and retina. DHA deficiency is associated with cognitive decline and increase neural cell death. DHA is depleted in the cerebral cortex of severely depressed patients.[124]
- **Gamma-linolenic acid,** an omega-6 fatty acid found in vegetable oil and was first extracted from the evening primrose. It is sold as a dietary supplement for treating problems with inflammation and auto-immune diseases. Research is ongoing on its therapeutic value for cancer to suppress tumor growth and metastasis.[125]

Many species of algae are tolerant of wide variations in growing conditions. Some species are nearly blind to geography.

*Geographical distribution*

Algae grow all over the Earth, including under both ice caps. Their preferred environments are in damp places or water but algae are common on land as well as in aquatic environments. Soils, rocks, trees and ice contain dried algae cells, and many are still viable. Various algae types can grow in all kinds of water which makes them excellent for pollution control.

Terrestrial algae have adapted to life on land and are commonly found in snow packs, deserts and grassland soils. These plants may be embedded in rocks found in deserts, mountains and even Antarctica. Rock algae, often combined with its symbiont lichen, enhances soil formation, water retention, increases availability of nutrients and minimizes soil erosion.[126]

Occasional algae blooms occur, usually due to an overabundance of nitrogen from agricultural waste streams and run-off. Blooms create high algal densities which may color the water and produce considerable $O_2$. Even with wind and wave action, the plants grow so quickly that trillions are shaded from sunshine and die.

**Algae Blooms**

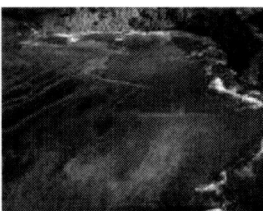

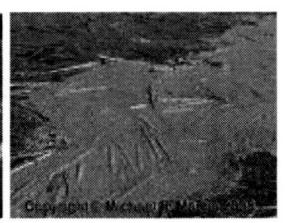

Fish and other sea creatures breathe oxygen similar to land animals but in lower quantifies. Fish absorb oxygen directly from the water into their bloodstream using gills while land animals use lungs to absorb oxygen from the atmosphere.

In an algae bloom the organic biomass becomes a feast for bacteria that work to decompose it and generate the pungent smell of hydrogen sulfide (rotten eggs) and ammonia. With plentiful food, bacteria multiply quickly and use all the dissolved oxygen in the water. When the dissolved oxygen content decreases, fish, aquatic insects and plants cannot survive. This results in dead zones in rivers, marshes and lakes as well as the Chesapeake Bay, Oregon Coast and the Gulf of Mexico.

# Chapter 5. Algae Growth and Production

Imagine the possibilities for food and fuel production from cultivating the fastest growing plant on Earth.

Algal production typically starts with the classic entrepreneurial question:

> What is the highest value product(s) that can be produced?

For decades that question has been answered based on local conditions where producers either harvested natural stands, enhanced natural settings or produced Spirulina in open ponds.

Considerable variation occurs in algal production due to the nature of the plant and the early stage of the industry. For example, the desired product characteristics should drive rational production decisions. In practice, some algae producers began growing algae because they:

- Had surplus tanks that spontaneously produced algae
- Found an open pond that grew algae naturally
- Found the aquaculture system could not grow striped bass in the summer but could grow algae in the heat
- Realized they could not continue to pay higher prices for feed corn for their fish so the grew algae as a substitute

- Had considerable experience with irrigation technologies and applied that knowledge to growing algae in tubes
- Read a "how-to" article and began producing algae[127]
- Decided to try an algae "home brew" for biodiesel

A host of other producers attacked production rationally and created fascinating business that are described in Chapter 8. Algae are old plants which is why they are superbly adept at proliferation.

### *Propagation*

Asexual algal propagation uses one or a combination of three strategies: cellular division, fragmentation and spores. Cellular division divides the cell in two, bipartition, and the cells separate. Fragmentation occurs when pieces break off the parent and begin growing independently. The spore strategy creates zoospores or buds which break off and move away from the parent and begin growing – similar to yeast.

Sexual reproduction is more complex but incredibly efficient. Some species use multiple strategies, including both sexual and asexual in different settings or different growing conditions. In one day, a plant reproducing by bipartition, fragmentation or spores may produce several million decedents.

Some algae reproduce sexually, some asexually, while many combine both modes. In some green algae, the type of reproduction may be altered if changes occur in environmental conditions, such as lack of moisture, mixing or nutrients.

Algae producers use a process called **stressing** where environmental conditions are changed in the growing tanks in order to push algae to produce special characteristics. Stressing may cause algae to grow faster or to produce more or less oils, proteins or carbohydrates.

Algae survived the harsh environments of early Earth by evolving the ability to grow quickly when conditions were favorable. When conditions changed, the plant died but left cells or spores that were viable indefinitely, until good growing conditions re-occurred. Algae

## Growth and Production

also employ a carrier strategy where animals or birds, (e.g., giant sloths or flamingos) take plant fragments to other settings.

### *Biomass production*

Cultivated algae grow quickly and display continuous growth in sunshine where the biomass may double or triple daily.[128] Algae slow their growth on cloudy days and go into respiration phase of photosynthesis at night. Algae grow similar to other plants and generally grow faster with increasing sunshine, warmth and nutrients.

Arizona State University Polytechnic Laboratory for Algae Research and Biotechnology, LARB

Algae grow within the boundaries of the "law of the minimum." The plant grows quickly to the maximum it can until it hits a mineral, chemical, nutrient, light or temperature limitation. When the last of the limiting nutrient is absorbed, nitrogen for example, the plant stops growing until more of the minimum constraint becomes available.[129] The challenge for algae cultivation becomes insuring that sufficient nutrients are continuously available.

Naturally occurring algae are often nutrient limited because their environment lacks specific nutrients and the plant composition changes as a function of food available. For example, kelp may be sulfur limited and the plant tends to produce a full set of amino acids except those associated with sulfur, lowering the quality of the plant's protein. Algae also show seasonal variation in composition from temperature changes, nutrient access or other stressors.

Algae differ widely in the levels of chemical, light and temperature parameters that limit their growth. For example, some algae flourish in low pH water (high acid) while others prefer high pH. Laboratory studies have determined optimal concentrations of nutrients and other growing parameters. Nutrient concentration ratios such as N/P combined with pH can predict which specific algal strains should predominate under stable resource conditions.

Most algae are not grazers; they have no propulsion, so they must rely on food that comes to them. Since the biomass grows so quickly and becomes so densely populated, mixing nutrients and sunshine so that all cells have access is a major challenge.

## Nutrients

Algae are fed fertilizers similar to land plants and need considerable $CO_2$ to grow productively. Growth medium recipes for each specific strain are available at the University of Texas algae culture collection site, http://web.biosci.utexas.edu/utex. The recipe for growing fresh water, blue-green algae Spirulina, may use the Allen medium.

**Table 5.1. Nutrient Recipe for Spirulina**

| Component | Amount | Stock solution concentration |
|---|---|---|
| 1. HEPES buffer | 2.3 g/L | |
| 2. NaNO$_3$ | 1.5 g/L | |
| 3. P-IV Metal Sol | 1 mL/L | |
| 4. K$_2$HPO$_4$ | 5 mL/L | 1.5 g/200 mL dH2O |
| 5. MgSO$_4$·7H$_2$O | 5 mL/L | 1.5 g/200 mL dH2O |
| 6. Na$_2$CO$_3$ | 5 mL/L | 8 g/200 mL dH2O |
| 7. CaCl$_2$·2H$_2$O | 10 mL/L | 5 g/200 mL dH2O |
| 8. NaNO$_3$ | 10 mL/L | 1.16 g/200 mL dH2O |
| 9. Citric Acid·H$_2$O | 1 mL/L | 1.2 g/200 mL dH2O |

Growth and Production

**Allen Medium.** Suitable for cultures of LB 1928 Spirulina platensis may include enriched seawater. For each liter:

1. To approximately 950 mL of $dH_2O$, (distilled water), add each of the components in order while stirring continuously.
2. Adjust pH to 7.8.
3. Bring total volume to one liter with $dH_2O$.
4. For 1% agar medium: add 10 g of agar to the flask; do not mix.
5. Convert and autoclave (heat) medium, then store at refrigerator temperature.

Substantial variation in growth medium recipes occurs for various strains and species. Producers also sometimes stress algae at a certain time in their growth cycle by adding or subtracting a specific nutrient. Stressing algal cells causes defense strategies such as faster or slower growth, storing more lipids or creating novel compounds. Algal cells use elements in many ways as illustrated in Table 5.2. [130]

### Table 5.2. Use of Elements in Algal Cells

| Element | Function and location in algal cells |
|---|---|
| Nitrogen | Amino acids come in nucleotides, chlorophyll, phycobilins |
| Phosphorus | ATP, DNA, phospholipids |
| Chlorine | Oxygen production in photosynthesis, trichloroethylene, perchlorolethylene |
| Sulfur | Amino acids, nitrogenase, thylakoid lipids, CoA, carry DN, agar, DMSP, biotin |
| Silicon | Diatom frustules, silicoflagellate skeletons, synurophyte scales, stomatocyt walls |
| Sodium | Nitrate reductase |
| Magnesium | Chlorophyll |

| | |
|---|---|
| Iron | Ferredoxin, chtochromes, nitrogenase, nitrates and nitrite reductase, catalase |
| Potassium | Agar and carrageen, osmotic regulation, cofactor for many enzymes |
| Molybdenum | Nitrate reductase, nitrogenase |
| Manganese | Oxygen evolving complex of photosystem II, cell walls |
| Zinc | Carbonic anhydrase, Cu/Cn superoxide dismutase, alcohol dehydrogenase, glutamic dehydrogenase |
| Cobalt | Vitamin B12 |

Some elements are provided to algae in trace amounts because very little is needed by the cells. When algae are stressed by limiting a nutrient, their defense mechanism activates and they begin storing the energy they need to survive.

### Nutrient stress

Algae's ability to proliferate over a wide range of environmental conditions reflects their diversity and occurs partially due to their ability to modify lipid metabolism efficiently in responses to changes in the environment conditions.[131] When the human body is stressed from fasting, cells shift into survival mode, revving up repair mechanisms and protective processes such as nutrient storage. Algae do the same thing and often increase lipid storage.

Under optimal conditions for growth, algae synthesize fatty acids principally for synthesis into glycerol-based membrane lipids which constitute 5 – 20% of their dry cell weight. Fatty acid esters form the backbone of DNA structures and are used by cells for the synthesis of membranes and other materials.

Algae fatty acids are enriched in the chloroplast and may be 14-, 16- or 18-carbon fatty acids or combinations. Fatty acids are used in the synthesis of cellular membranes and lipids. Many algal species have been found to grow rapidly and produce substantial amounts of oil and are referred to as oleaginous algae. Oleaginous algae can produce lipids up to 28 times faster per unit area than the best land plants.[132]

Under unfavorable environmental or stress conditions for growth, algae may alter their biosynthetic pathways and accumulate neutral lipids up to 50% of their dry cell weight, mainly in the form of triacylglycerol called TAG. Unlike membrane lipids, TAGs do not perform a structural role but instead serve primarily as a storage form of carbon and energy. TAGs are the lipids used for biodiesel.

In higher plants, individual classes of lipids may be synthesized in localized specific cells, tissues or organs such as a seeds or fruit. Algae assimilate several types of lipids and they do it in a single algal cell. After being synthesized, TAGs are deposited as densely packed lipid bodies located in the cytoplasm of the algal cell.[133]

Oleaginous green algae display about 25% dry cell weight under normal growing conditions but can store up to 50% under stress conditions. The intrinsic ability to produce large quantities of lipid is species and strain specific rather than genus specific.[134] For example, green algae and diatoms tend to double their lipid storage under conditions of stress but many blue-green algae or cyanobacteria do not accumulate lipids under stress.

Algae synthesize fatty acids as building blocks for the formation of various types of lipids. Fatty acids are either saturated or unsaturated and may be medium, long and very long chain. The quality and energy produced by biodiesel produced from algae is determined largely by the structure of the component fatty acid esters.[135]

Chemical and physical mechanisms are used to stimulate changes in lipids and fatty acid composition in algal cells. Chemical stimuli include nutrient starvation, salinity and growth medium pH. Physical stimuli include temperature and light intensity.

Nitrogen limitation is the single most critical nutrient affecting lipid metabolism and algae. Deficiencies of other nutrients, especially phosphate and sulfates, also promote lipid accumulation, although not as much as nitrogen.

Temperature has a major effect on fatty acid composition as decreasing temperature is increase fatty acid unsaturation. As temperatures increase algae tend to produce more saturated fatty acids.[136] Increasing temperature tends to increase total lipid content.

Variations in light intensity cause changes in algal chemical composition, pigments and photosynthetic activity. Low light intensity induces polar lipids while high light intensity increases neutral storage lipids, mainly TAGs.[137] Other factors that induce composition changes in fatty acids include position in the growth cycle when stress is induced, especially the stationary phase, and culture age or senescence.[138]

## Cultivation

Algae grow in open, closed or semi-closed systems in round, long or tubular tanks that maximize access of the entire biomass to sunlight. Growth occurs only in the top layer, about two inches, of the growing medium, usually water. New cell growth blocks the sunlight for plants below. Semi-continuous mixing is necessary to give all the algae sufficient light.[139] Some production systems put light sources in the water to augment sunlight.

Growth occurs based on a host of variables that not only constrain growth but may change the algal composition. Primary variables are:

- **Light.** Usually sunlight provides sufficient light but artificial light also works as well – especially for indoor growing systems. Some growing systems may be tilted to optimize orientation to the sun and reflected light.
- **Mixing.** Since most growth takes place in the top layer of the surface that faces the light source, mixing is imperative. Each cell needs to move in and out of the light for their light and dark growth periods as they take in $CO_2$ and exhale $O_2$. Algae are heavier than water would sink away from their light source without mixing.
- Algae grow so fast they become **nutrient limited** quickly in still water. They cannot move and graze for food because they usually have no propulsion. Mixing brings nutrients and

especially $CO_2$ to each algae cell. Mixing also helps release $O_2$ from the water to the atmosphere. Too much or too little mixing impedes growth and rough mixing methods may create cell damage from shear stress.[140]
- Some algae have evolved two interesting differentiated features: **flagella and eye spots**. At a specific growth stage, some algae grow flagella, slender projections from the body like sperm tails that move in a whip-like motion to propel the algae. The eye spot recognizes light and the flagella propel the plant toward the light. Movement is very slow, possibly an inch an hour.
- **Water.** Algae grow well in nearly any kind of water. They are especially good at using photosynthesis to convert dissolved nutrients and metals in waste water to green biomass where the metals can be removed and recovered.[141] Production systems can use wastewater, grey water, and saline or ocean water, depending on the species grown. Growing systems can recycle the water so the only loss comes from evaporation.[142]
- $CO_2$. Algae's favorite food, $CO_2$, needs to be added as a gas or in bicarbonate form because cultivated algae grow too fast to be able to take sufficient $CO_2$ from the air. Some manufacturers and coal-fired power plants flue their emissions through algal ponds which convert the heat and $CO_2$ to algal biomass and $O_2$.[143]
- **Nutrients.** Algae feed their growth with the same fertilizers used for land plants.[144] Growth requires far less nitrogen and other fertilizers per pound of biomass than corn and the dissolved nutrients are easier and less expensive to apply. Dissolved fertilizer is utilized with far more efficiency than land plants. Unused fertilizer can be reused with the recycled water.
- **pH.** The acidity of water may be specific to the type of algae produced. Controlling the water's pH represents a good strategy for retarding growth of competing algae. Water pH is likely to be highest at noon due to the high photosynthetic activity which consumes maximum $CO_2$.[145]
- **Stability.** Maintaining a stable growth environment presents difficulties with the high velocity of growth. The growing medium

may retain too much of any nutrient or $O_2$ which may create stress and growth or composition changes on the plants.[146]

Power companies such as Arizona Public Service have turned their problem with $CO_2$ emissions into an opportunity. The APS Redhawk 1,040 megawatt power plant recycles greenhouse gases into renewable biofuels and uses algae to capture the $CO_2$ gas emissions. The power plant exhaust is routed through algae growing systems which remove part of their $CO_2$ emissions during the day.[147] Power plants run 24/7, so this presents only a partial solution. In addition, power plant $CO_2$ emissions are so large that current technology allows uptake of only about 20% of available $CO_2$.

Some power plants also use waste heat from power generation in the growing systems that increase the velocity of biomass growth. The only company supplying these systems currently, Greenfuels Technologies, claims that using algae-fed $CO_2$ and warm water from the power plant can potentially create annual yields of 8,000 gallons of biodiesel plus about 8,000 gallons of ethanol per acre.[148] These production levels may be theoretically possible but are well beyond current operational systems.

### *Algaculture production systems*

Algal biomass grows in ponds, tanks or tubes called biofactories or algaculture production systems. Water, inorganic nutrients, $CO_2$ and light is provided to the microalgal culture during biomass growth, Figure 5.2.

Algae prefer diffused light that is not too bright so some systems use shading that both limits light and diffuses it. Various species produce best at different temperatures so some systems use recycled water on the outside of the biofactory to maintain optimum temperature.

Even though $CO_2$ may be about 2% of production cost, that cost can be minimized by siting the biofactory near a power or manufacturing plant the produces $CO_2$. Nutrients may be provided from wastewater, recovered from the algal tank or harvested fertilizer. After the algal oil is removed, the remaining biomass contains considerable nutrients.

## Growth and Production

**Figure 5.2. Biomass growth**

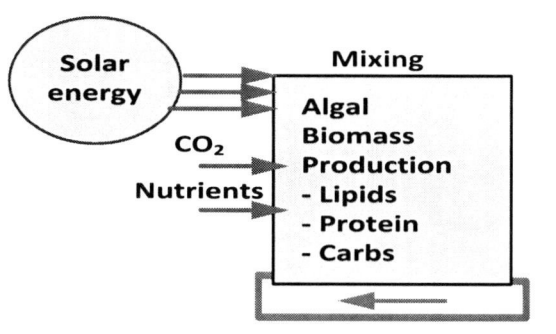

Solar energy → Mixing
$CO_2$ →
Nutrients →
Algal Biomass Production
- Lipids
- Protein
- Carbs

← Recycle water and nutrients

Closed systems offer the advantage that high nutrient water may be recycled through the system. This practice significantly lowers the cost of added nutrients. It also minimizes water loss to evaporation.

Algaculture systems that use high-saline water, such as agricultural waste streams or brine water, produce a biomass with considerable salt that must be removed before it can be used as fertilizer.

Harvest may occur daily by filtering, centrifuge or flocculation, Figure 5.3. The cells suspended in the broth are separated from the water and residual nutrients are recycled to biomass production. Algal oil is extracted from the recovered biomass converted to biodiesel. Some of the non-oil biomass may be used as animal feed, fertilizer and for other coproducts.

Part of the spent biomass undergoes anaerobic digestion to produce biogas that generates electricity which powers the biomass mixing and water transport.[149] Effluents from anaerobic digestion may be used as a nutrient-rich fertilizer for more algae production or as nitrogen rich irrigation water. Most of the power generated from the biogas is consumed in biomass-production and any excess energy may be sold to grid. Some systems use solar panels with photovoltaic cells to convert solar energy directly to electricity which is either used directly, used to warm production water or store in batteries.

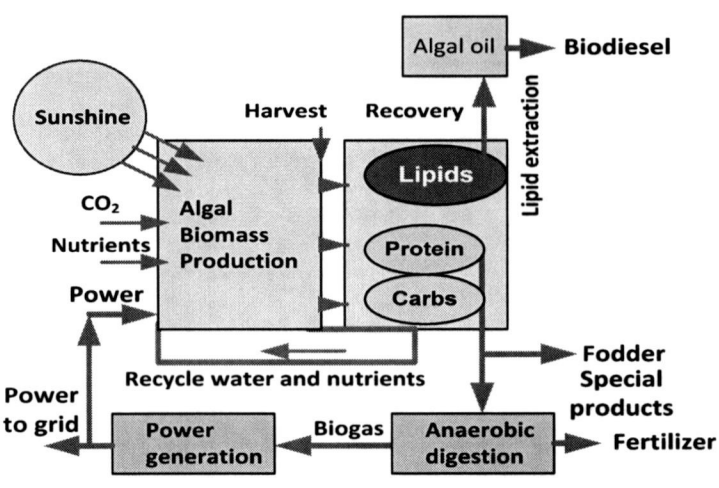

## Figure 5.3. Algaculture production system

Approximately half of the microalgal biomass dry weight is carbon, typically derived from $CO_2$ or carbonates, and is fed continually during daylight. Each 100 tons of algal biomass fixes roughly 183 tons of $CO_2$.

Production water may be too salty for land plants, to acidic (wastewater) or contain dangerous chemicals from industrial effluent that the algae clean from the water. The chemicals may be separated safely during processing although such algaculture systems produce primarily energy feedstock rather than food. The non-harvested algal biomass seeds the next production cycle, which may be the next day.

In a continuous culture, fresh culture medium is fed at a constant rate and the same quantity of microalgal broth is withdrawn. Feeding stops during the night but mixing continues to prevent biomass settling. As much as 20% of the biomass produced during daylight may be consumed during the night to sustain the cells until sunrise.[150] Nightly biomass loss depends on the growth light level, growth temperature and the temperature at night. Some production systems are experimenting with night lights to boost productivity.

Biomass composition varies by variety but may be 50:25, oil to protein, with about 15% carbohydrates and 10% ash or waste.[151]. After the oil component is used for biofuel, the remaining high protein biomass may be demoistured and stored in a convenient form such as a cake which does not require refrigeration and has about a two year shelf life. The algal cake may be separated into various food, food ingredients, fodder, fertilizer, fine medicines or other components.

Algal biodiesel production is carbon neutral because the power needed for producing and processing the algae can come from the methane produced by anaerobic digestion of the biomass residue remaining after oil extraction. The modest energy requirement may also come from other non-carbon sources such as solar.

Typical components are processed into a variety of products, Figure 5.4, illustrating a highly flexible source for fuels and foods. The potential product mix offers almost infinite variations, depending on algae strain and production parameters.

The harvested biomass is extremely malleable in the sense that it can be stored in the same form as corn, wheat, rice or soy products. These include protein-rich milk, soft mash of any size, shape or texture, tortilla, cracker or flour. The biomass may be made into texturized vegetable protein with added fiber or extruded to make additives for meats that improve moisture retention and increase protein while lowering fats.

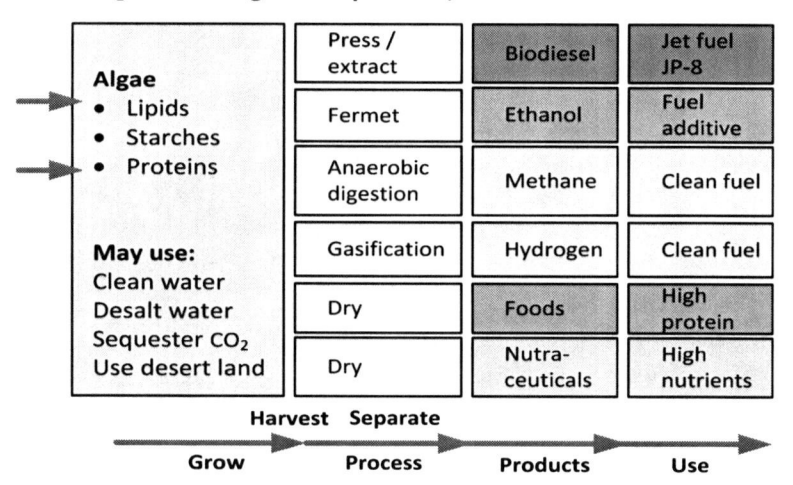

### Figure 5.4. Algae Components, Products and Uses

| Algae<br>• Lipids<br>• Starches<br>• Proteins | Press / extract | Biodiesel | Jet fuel JP-8 |
|---|---|---|---|
| | Fermet | Ethanol | Fuel additive |
| | Anaerobic digestion | Methane | Clean fuel |
| May use:<br>Clean water<br>Desalt water<br>Sequester CO$_2$<br>Use desert land | Gasification | Hydrogen | Clean fuel |
| | Dry | Foods | High protein |
| | Dry | Nutra-ceuticals | High nutrients |

| Grow | Harvest<br>Process | Separate<br>Products | Use |

Processing can match the form of nearly any food such as peanuts, pasta, pesto or protein bars. Fortunately, years of food processing for land-based plants that have an unappealing natural taste such as soybeans make it easy to add flavors, textures (fibers) and aromas.

The desired product outcome drives strain selection, growth parameters and processing. Better algal strains through screening natural varieties, hybridization and bio-engineering may motivate some growing systems to be dedicated to one or both foods or fuels.

An understanding of this plant's characteristics explains its productivity advantage for creating food and fuel biomass. A set of production and value comparisons between land and water-based plants; corn and algae is shown in Table 5.3.

### Table 5.3. Production Comparison – Corn and Algae

| Requirements | Corn – Land-Based | Algae – Water-Based |
|---|---|---|
| Time requirements | High – full growing season, 90 – 120 days | Low  One day |
| Weather requirements | Storms, drought cold or heat kills crop  $65°$ F to $95°$ F; 120 days | Too cold or no sun slows growth  $65°$ F to $115°$ F; 1 day |
| Risk of crop failure | High – weather, drought or pests | Low  Controlled environment |
| Soil requirements | High. Good soil, tillable, holds moisture, drains | Low. Algaculture systems may sit on desert or rooftops |
| Fossil fuel use per unit of oil | High (7 units)  Diesel, gas and coal | Low (1 unit)  Solar or electrical |
| Fertilizer requirements | High  Fertilize in the field  May waste 50% | Moderate  Dissolves in water  No waste due to reuse |

## Growth and Production

| | | |
|---|---|---|
| **Herbicide requirements** | High – must control competing weeds | Low<br>None |
| **Pesticide requirements** | High<br>Must control pests | Moderate<br>Few pests |
| **Run-off** | Significant<br>Rain or irrigation | None<br>Controlled environment |
| **Equipment requirements** | High<br>Heavy investment to grow and harvest corn | Moderate to high<br>Algaculture systems are expensive to build and maintain |
| **Physical labor** | Considerable<br>Tilling, planting, fertilizing, harvesting<br>Little work in winter | Moderate<br>Automated growing and harvesting Continuous, all year |
| **Health risk** | High – accidents and ag chemicals | Minimal – no heavy equipment or chemicals exposure |
| **Tilling soil** | Significant<br>Even no-till requires tilling | None<br>Avoids dust and erosion |
| **Heavy equipment** | Significant<br>Growing, harvesting, transporting | None<br>Light vehicles only |
| **Growing region** | Limited<br>Narrow latitude<br>Good soils, water | Broad<br>Needs sunlight |

Cultivated algae require minimal land and water footprints and both land and water can be of such poor quality that they are not useable for land-based food plants.

Algae cultivation typically occurs in tanks or ponds, so no soil tilling, heavy equipment or pesticides and herbicides are required, although light tractors are common. Algae grow all over the Earth, so its range substantially exceeds corn. However, cultivated algae grow best in sunny, warm regions. Algae can grow where other crops cannot grow, such as deserts, mountains and rooftops.

Algae do not have the cellulosic trunk, tassel, leaves, roots and seeds – the structural overhead – necessary for land plants to withstand the land environment. Algae invest their growth energy in creating oils and proteins with light carbohydrates for the cell walls. An algae strain with 60% lipids produces over 50% net oils that can be made into liquid fuel like high-powered jet fuel or biodiesel.

Table 5.4. Production Comparison – Value

| Value | Corn – Land-Based | Algae – Water-Based |
|---|---|---|
| Biomass reliability | **Moderate** Threats: weather, water, weeds, mildew, diseases and pests | **High** Threats: susceptible to contamination, fowling and disease |
| Biomass oil productivity | Low – 3% 97% non-oil biomass with cellulosic structure | High – 60% Plant packed with oils, carbs and protein |
| Protein value | Low – 2% of plant Low – 34% of kernel Most of the plant is cellulose | High – 30% Protein varies 20% to 60% depending on strain |
| Bioavailability – access to oils, protein | Low Hard cellulosic structure | Medium Some algal strains have soft cell walls |
| Plant waste | High – plant mostly wasted biomass with cellulosic structure | 10% ash – Plant packed with protein, carbs and oils |

Growth and Production

*Productivity metrics*

Biomass production yields and parameters, such as the water required, have been developed and tested at the Laboratory for Algal research on Biofuels, LARB, at Arizona State University directed by Professors Qiang Hu and Milton Sommerfeld.

The LARB field laboratory is not fully scaled to an acre but produces the equivalent of 6,300 gallons of oil per acre. The LARB production uses a 330-day production year with 30 days of scheduled maintenance and five days of algae rest – predictable cloudy weather in central Arizona. The **Sustainable Biofuels Scorecard** illustrates the differences between corn and algae in Table 5.5.

Table 5.5. Algal productivity compared with Corn

| Parameter | Corn | Algae | Advantage |
|---|---|---|---|
| Water footprint | 1,000,000 gal per acre if irrigated* | 10,000 gal/acre (evaporation) Wastewater, brackish, saline | 1/1000$^{th}$ 0.001 May reuse water |
| Earth footprint for 4 M gal | 10,000 acres* Good cropland | 450 acres desert, rocky, non-crop soils | 1/22th** Non-cropland |
| High energy yields | 350 gal / acre* = 224 gasoline equivalent gal | 6,300 gal / acre = 6,300 GEG | 28 times |
| High energy fuels | Low Ethanol 64% of gasoline | High Jet fuel – JP-8 Green diesel, biodiesel, Hydrogen | 34% - 50% energy advantage Ethanol/ biodiesel |
| Net energy value | 0* Equal energy input and output | + 60% 10% to grow 30% to refine | 60% NEV advantage |

| | | | |
|---|---|---|---|
| **Small carbon footprint** | Poor<br>CO$_2$ emissions | Excellent<br>Gives off O$_2$ | Carbon advantage<br>Sequesters CO$_2$ |
| **Ecological footprint** | Poor<br>Pollutes soil, water and air | Excellent<br>No pollution | Ecologically clean |
| **Net food yield** | Modest<br>Distillers' grains only; cattle feed that may be extra cost to dry and ship | Good<br>Protein + carbohydrate; human or animal food, nutrients and medicines | Net food advantage<br>Also specific nutrients, pigments and medicines. |
| **Avoid mono-cropping** | Poor<br>Requires monocropping | Excellent<br>May grow a wide variety of plants | Monocropping advantage |
| **Economically sensible** | Poor<br>Requires huge subsidies | Unknown<br>Will require subsidies | Unknown<br>No large sites |

\* Ignores crop rotation requirement. True disadvantage is double.

\*\* The DOE / NREL *Algal species Report* shows 30 times higher land efficiency or 43 acres to produce 4 million gallons of oil.[152]

The 28 times oil production advantage represents a significant oil productivity advantage. Not only is the biomass production advantage extraordinary but the other key parameters also favor algae.

***Are these numbers real?***

The magnitude of biomass productivity and other characteristics may seem an impossible exaggeration. They are not. These parameters represent a set of extraordinary productivity variations that differentiate traditional agriculture and algaculture growing water-based plants. Many of these differences have been known for decades.

The comparisons used here are based on conservative biomass production for existing growing systems.[153] The literature on water-based plant production is filled with theoretical exaggerations from optimistic people who project a short lab-based yield to commercial production. Real-world, large-scale production creates a number of inefficiencies that currently reduce theoretical biomass production by at least a factor of 10. The good news: even dividing by 10 creates significant productivity advantages versus land plants.

Existing growing systems can be improved substantially by selecting more vigorous algal strains such as strains with higher oils or protein, better production facilities and advances in biotechnology.

*Hydrogen production*

NREL and other labs are conducting R&D on algae as a biological hydrogen source. In 1939, a German researcher named Hans Gaffron at the University of Chicago, observed the green algae could switch from the production of oxygen to the production of hydrogen.

Anastasios Melis, a researcher at the University of California at Berkeley discovered in 1998 that by depriving algae of sulfur it would switch from its normal behavior where it produced oxygen to the production of hydrogen. He found that the enzyme responsible for this reaction is hydrogenase but that the hydrogenase will not cause this switch in the presence of oxygen.

Melis found that depleting the amount of sulfur available to the algae interrupted its internal oxygen flow, allowing the hydrogenase an environment in which it can react, causing the algae to produce hydrogen. Unfortunately, production methods to date have produced only minuscule amounts of hydrogen from algae but hydrogen production remains a very active area of research.

Algae have been attacked by herbivores for several billion years longer than land plants. Algae employ an astonishing array of defense strategies, especially their ability to quickly produce chemical compounds. Most algal compounds are non-toxic. Some algae are epibionts and spend most of their life attached to another organism. Various species live on protozoa, zooplankton or larger animals such as turtles, birds, whales, giant sloths and polar bears. These associations may be mutually beneficial as algae shield the host from over exposure to the sun, provide food or serve a cleaning function.

One defense strategy that dominates people's knowledge of algae, killer tides, occurs from a few single-celled organisms called dinoflagellates. Dinoflagellates may produce deadly toxins that create red tides and can multiply rapidly because they can fix nitrogen.

**Red Tide**

A red tide may cause massive fish kills when toxins overwhelm fish. Red tides can cause eye and skin irritations for people. Red tides have killed fish off California, Cape Cod, Gulf of Mexico, Peru, Japan, Australia, Africa and the Mediterranean.

Shellfish, such as clams, scallops, and mussels, are filter feeders. They consume plankton (dinoflagellates) and filter the cells out of the water. Shellfish concentrate toxin in a special organ but are immune to the poison. When humans eat the poisoned shellfish, about an hour later the poison affects the nervous system and the victim experiences a numbing of the lips, tongue, and fingertips. Respiratory failure may occur if the patient is not kept alive by artificial respiration until the effects of the toxins pass.

Algae pollution, usually from blue-green algae, cyanobacteria, creates difficult problems for water purification. These single celled organisms are so small they are difficult to remove with standard filtering technologies. Fortunately, several new technologies using algae as living censors are providing both test and resolution for the rare occurrence of toxins.[154]

# Chapter 6. Algal Production Challenges

> History is littered with the debris of projects that were theoretically possible but economically impractical.
> 
> Warren Belasco, *Meals to Come: A History of the Future of Food.*

> If the production of algae and recovery of constituent products were easy, it would have happed 50 years ago. Commercial scale production awaits R3D – research and development, demonstration and diffusion.

Scalability presents the primary production challenge. Laboratory conditions that enable algae to grow many times more productively than land plants simply have not been realized in field settings.

Experiments in modestly scaled production have consistently shown that in field settings the biomass cannot withstand ambient temperatures, are inconsistent in production, too easily become unstable and simply stop growing. Fast growing pure algal cultures do not remain clean indefinitely, and "weed" algae must be removed. Field settings, especially open ponds, do not allow growers to control:

- **Temperature** – A temperature extreme, either high or low, can damage or destroy an open algal pond. Closed algaculture systems allow growers to stream water over the outside of glass on hot days to moderate hot temperatures. Producers can simply stop growing algae in severely cold weather.

- **Opportunistic invasion** – Winds, birds, insects, fungi and other vectors bring new algal strains to growing ponds. Survival of the fittest operates in nature and a high lipids algal species may be replaced by a species with low oils. More commonly, an algal pond becomes a potpourri of different species making harvesting and extraction of potentially valuable components difficult.
- **Mixing** – Ponds limit mixing alternatives and often have a single paddlewheel that moves the algae slowly around an oval track. Algae have a dark and light cycle and grow best when they have for example, access to light every 10 seconds. Mixing solutions in algaculture systems enable precise mixing velocities that are impractical in ponds.
- **Nutrient delivery and access** – Nutrients are an important limit to growth for algae and ponds make it difficult for consistent and timely nutrient delivery to the growing biomass. When algae do not have access to sufficient nutrients, the biomass simply halts growth or stores available nutrients.

### Algal ponds and a Plastic Bag

Added to the challenges of field settings is cost because in order to be commercially viable, algal production must occur at lower dollar and energy costs than energy alternatives. The NREL Algal species Program, for example, concluded that closed systems were impractical for algal production because they were too expensive to build and maintain. Unsurprisingly, nearly all algal production to date occurs in open ponds.

Elements of algal production shown in Figure 6.1 include growing systems, inputs, processing and marketing.

Production Challenges

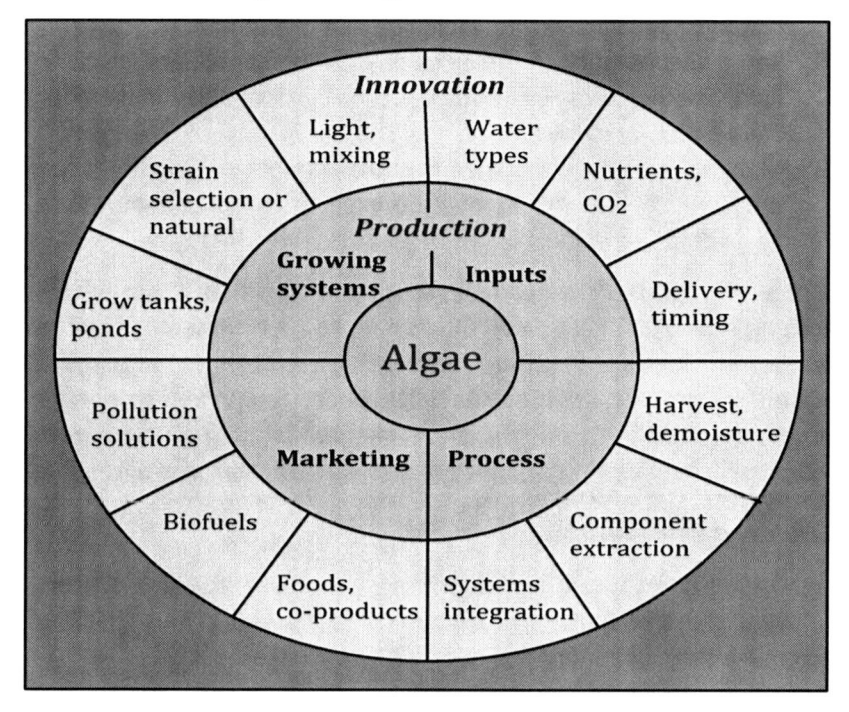

**Figure 6.1 Algal Production R&D**

[Figure 6.1: A circular diagram centered on "Algae". Inner ring segments: Production (Growing systems, Inputs), Process, Marketing. Outer ring segments labeled: Innovation, Light/mixing, Water types, Nutrients/CO₂, Delivery/timing, Harvest/demoisture, Component extraction, Systems integration, Foods/co-products, Biofuels, Pollution solutions, Grow tanks/ponds, Strain selection or natural.]

Years of algal research have produced the following general insights on algal production.

1. Each species reacts differently to variations in growing parameters and each strain within a species may have different reactions to variations in growing parameters.
2. Stressing algae by withholding nutrients or changing other growing parameters causes the plant to implement a defense strategy for survival and to change the biomass composition.
3. The optimal temperature range for growing algae is 55° to 95° F, although some species can grow at almost any temperature above zero. Most species have a maximum productivity sweet spot and slow growth outside their favored temperature range.
4. Production is optimized with intermittent light and dark cycles of five to 15 seconds. Container shape, mixing and light sources determine light cycles.

5. Most species are vigorous in water with a wide range of dissolved salts and waste, including heavy metals.
6. Some species stick to the sides of growing containers, blocking light to other cells while others do not. Sticky algae make poor candidates for cultivation.
7. Many species have a relatively narrow acceptable acidity range measured by pH. Acidity may be used to control opportunistic algae contamination.

Possibly the most critical variable in algal production is not visible because the algal cells are too small to see except under a microscope. Species or strain selection are critical to algaculture because species are selected that are robust (grow well in a wide variety of conditions) and maximize the production of the target product which may be lipids, protein, cell walls or other attribute.

### *Species and strain selection*

Algae offer so many product alternatives that eventually producers will use a checklist that may look similar to Table 6.1. This checklist also answers the question: "What do algae produce?"

#### Table 6.1 Algal Characteristics Selection

| Characteristic | Threshold | Characteristic | Threshold |
|---|---|---|---|
| Lipids | > 60% | Calcium | > 5% |
| Proteins | > 30% | Boron | > 5% |
| Enzymes | Specified | Other | > 5% |
| Antibodies | Specified | Soft cell walls | 4 out of 10 |
| Vaccines | Specified | **Pigments** | |
| Optimal mix of lipids / protein | L > 30%<br>P > 30% | B-Carotene | > 1% |
| Carbohydrates | < 10% | Lutein | > 1% |

Production Challenges

| Vitamins | | Pharmaceuticals | |
|---|---|---|---|
| B, C, D, E | > 1% | Antibiotics | yes |
| Polysaccharides | | Anti-tumor/cancer | yes |
| Agarose | yes | Anti-HIV substances | yes |
| Agaropectin | yes | Antivirals | yes |
| Sodium alginates | yes | Designer drugs | yes |
| Sulfated polys | yes | Nutraceuticals | yes |
| Destrin | yes | Polyunsaturat. fatty | |
| Carrageenans | yes | Eicosapentaeocic | yes |
| Minerals | | Docosahexaenic | present |
| Zinc | > 5% | Arachidonic acid | yes |
| Iron | > 5% | Other | |
| Selenium | > 5% | Other | |

Strain selection also considers species productivity. For example, for 50 years an algal species has been known that produces around 82% lipids. The strain is seldom selected for use because it grows so slowly.

The University of Texas maintains a list of 3,000 living algal species that can be sorted based on characteristic, e.g. high algae oils. UTEX provides algae cultures at modest cost for research, teaching, biotechnology development and various other projects throughout the world. Their website lists the cultures maintained by UTEX, conditions for their long-term growth and information regarding the purchase of cultures.[155] The UTEX web site also offers specialty strains such as algae that grow in freshwater, extreme environments, snow and salt plains. Extreme environment strains come from tough settings such as Antarctica and the Gobi Desert. After strain selection, the next big decision analyzes the growing system design.

## Green Algae Strategy

### *Green solar systems*

Green solar production systems are designed to capture maximum sunshine. Growing containers provide considerable visual variety and may be ponds, plastic bags, plastic sheets, resins or glass – anything that allows light to penetrate. Growing systems may be large medium or small and use natural sunlight or artificial light.[156] Some systems use fiber optics or mirrors for additional light.

Green solar systems are commonly called photobioreactors which our consumer research indicates creates fear and negative feelings in the minds of consumers. Even though the term photobioreactors implies the sun excites plant cells to produce biomass through photosynthesis, naïve observer's associate reactors with nuclear power. Additionally, the term bioreactor has become synonymous with garbage waste disposal.[157] Consequently, the terms used here are biofactory, green solar and algaculture production systems.

Green solar systems vary from uncontrolled settings in the ocean to semi-controlled settings in estuaries, lakes, rivers, wetlands and ponds. Controlled and semi-controlled growing systems may be made of any material that allows light to pass. Containers must hold water and may be in shapes such as tubes, rectangles, barrels, blatters or bags. Hybrid systems may start algae in the controlled environment of tanks and grow the production biomass in ponds.

### Rectangular Biofactories

Growers have traditionally used open ponds because they are easy to build and operate. Open ponds are inexpensive growing systems but allow algal species contamination, predator invasions, loss of water due to evaporation and other forms of contamination. Ponds enable opportunistic algal species to contaminate the pond and possibly dominate production based on growing conditions. An algal pond may be infested with ciliates, amoeba or flagellates which can decimate the algal biomass within hours. An open algal pond loses about as much water due to evaporation as a grain field consumes in irrigation. Ponds may become contaminated with insects, chemicals, disease microbes, heavy metals or weedy algae such as toxic cyanobacteria.

Closed algaculture systems producers select algal species that maximize the characteristics desired such as biomass percentage of lipids, protein, or component product. Food production would select to maximize biomass protein while biofuels may select a species with high lipid content.

**Vertical, angled or horizontal.** Algae are solar collectors so the plants benefit from maximum exposure to the sun. Some angled green solar systems track the sun similar to photovoltaic solar collectors.

Horizontal algaculture systems, typically tubular or plastic bags, provide another variation in solar exposure. Several companies use plastic bags that are typically rectangular or oval. XL Industries, applies irrigation technology to algal production and their XL Trough uses a specially designed tube laying system that rolls out 10 tubes in furrows. Their tubular system is simple and inexpensive.

### Tubular Biofactories

**Rectangular or tubular?** Different shapes provide different levels of solar exposure. A wide rectangle, similar to an aquarium, holds a lot of water but does not allow each alga cell to have sun exposure very often. Consequently, thin rectangular tanks, about three inches thick, tend to out produce tanks that are wider. Tubular tanks may be a few inches wider because they present more surface area around the circumference. However, tubes around 6 inches typically out produce tubes that are wider.

## Table 6.2 Algaculture Production System Trade-offs

| Type | Description | Limitations |
|---|---|---|
| Open pond | Economical, easy to manage, good for mass cultivation of algae, considerable global experience | Low culture control, Stability issues, Weak productivity, High land use, Species contamination, Evaporation problems |
| Vertical column | High mass transfer, good mixing with low shear stress, low energy consumption, scalable | Small illumination surface, Expensive construction, Shear stress problems, Cleaning issues |
| Flat rectangle | Large illumination surface area, good light path, good biomass productivities, relatively cheap, easy to clean, low oxygen build-up | Scale-up challenges, Culture stability, Temperature stability, Possible shear stress |
| Tubular | Large illumination surface, good light path, relatively cheap | Gradients of pH, dissolved oxygen and $CO_2$ along tubes, fouling, high land use |

The algal industry will continue to experiment with variations in biofactory shapes, sizes and with open and closed systems. It seems logical that low cost, low output systems will use open systems while those applications producing algal oil for biofuels will use closed systems to maximize growth speed, vitality and species homogeneity.

### *Light and mixing*

In natural settings, algae grow so fast that the new growth occurring closest to the sunlight tends to shade prior cells. Cells that cannot get light stop growing, stop producing biomass and stop propagating.

## Production Challenges

Adding to the complexity of shape, different algal species exhibit variation in their appetites for light and dark. In general, algae seem to grow most productively in a setting where they have access to light about every 10 seconds. During each brief dark period, algae seem to be digesting the light. This is why thin tanks tend to out-produce thick tanks. Some species are sensitive to the harsh light and grow best in soft light.

A few algae produce biomass in the dark and use sugar rather than sunshine for energy. Feeding algae sugar makes it difficult to achieve a net positive energy value due to the cost of producing sugar unless biomass growth is extremely rapid.

Theoretically, faster mixing in larger tanks should enable thicker tanks and high production. However, mixing above a relatively modest velocity threshold stresses and injures the soft plant cells and may create instability and the plants may stop propagating.

An algaculture system that receives only intermittent mixing produces biomass but at a slower rate. Therefore, constant, relatively gentle mixing aids productivity. The mixing process in ponds is commonly a paddlewheel on an oval raceway. The paddle wheel keeps the suspended algae moving. Bottom friction creates some turbulence away from the paddle wheel. In closed algaculture systems, compressed air and $CO_2$ is bubbled through the tank and the bubbles provide turbulence for mixing.

### *Water types*

Algae can grow in salt, brackish, human waste or industrial wastewater. Land plants die in saline water because the salt ions concentrate in the roots and block water transport through the plant. Algae have no problem with salt or heavy metals because they have no roots. Some species simply absorb the salt or heavy metals from the water leaving it clean enough for crop irrigation. After the algae are harvested, the salt, heavy metals or other pollutants may be recovered as coproducts.

Protein recovered from wastewater production systems meets food quality standards but may not meet consumer perceptions of cleanliness. Organic fertilizers used on terrestrial food crops are made from the components of wastewater --- primarily urea. The total growing process for food requires much higher standards of cleanliness than production systems optimized for oils or other coproducts. Production standards rise further if the producer grows food with an organic label.

## Nutrients and $CO_2$

Algae consume about the same nutrients in the form of fertilizers as land plants – only in smaller proportions. Since the biomass grows so fast to very high densities, the plants need nutrients regularly.

Corn for example, assimilates only about half of the nitrogen fertilizer put on a field. Algae may not absorb all the nitrogen fertilizer in a growth culture but the remainder may be recovered and recycled, along with the water for the next growing cycle.

Algae can absorb $CO_2$ from the water but water carries little $CO_2$, about 0.5% in solution. Most growing systems use supplemental $CO_2$ that is either bubbled into the tanks as a gas or added to the tank has a solid, typically a carbonate. Siting biofactories near $CO_2$ sources such as power plants or beer manufacturing plants saves input costs and sequesters $CO_2$.

## Delivery and timing

Nutrients are typically delivered into the growing medium either constantly or intermittently during the day when algae are growing. There is no need for nutrient delivery at night when algae are resting.

Some species like burst feeding where large amounts of nutrients are introduced at one time. More commonly, most algae prefer drip feeding where nutrients are presented in a fairly constant pattern throughout the day.

Algae display variations in the speed of growth at various times during the day. A common pattern is a steady increase in the speed of growth throughout the morning until about midday. The plants do not exactly take a siesta but they tend to slow their growth as the afternoon progresses. Growth slows in the early morning and late afternoon when the plants have less access to sunlight.

### *Harvest and demoisturing*

Considerable variation in harvesting reflects variations in species, size and available technology. The easiest harvesting method, settling, allows the culture to sit quietly overnight and the cells simply sink to the bottom where they can be removed and dried. A basin in sand covered with fabric allows the water to flow through to the sand leaving the algae to dry. It then can be scraped off the fabric.

Filtration may use fabric, cheese cloth or microscreen filters. In high productive systems, harvesting occurs once a day at maximum cell density which is typically late morning. A third to one half of the algal cells may be removed from the growing medium.

Some species are so small they require flocculation. Flocculation is derived from the word floc or flakes of material. When a solution is flocculated, the small solids are formed into clumps of aggregate which are easier to see and to remove with filters or screens. Flocculants are commonly used in water treatment to improve sedimentation and filterability of small particles. Alum, ferric chloride and Chitosin are common flocculants. Flocculation may be too expensive for large-scale algal production.

Froth flotation, another harvest method, aerates the water into froth and the algae are skimmed off the froth. Interrupting algae's carbon dioxide supply can cause algae to flocculate on its own, which is called autoflocculation.

After harvest, algae are scraped off the filter. In some cases, algae are dried in the sun similar to grapes that are dried to make raisins. Other production systems use a centrifuge to demoisturize the algae down to about 5% water. The extraction plan may be driven by convenience or available technology. Ideally, component separation enables the maximum number of components to be extracted from the biomass.

*Oil extraction*

Algal oil extraction methods vary from pressing to chemicals and sound. Algae vary widely in their physical attributes so various press configurations such as screw, expeller or piston are matched for the type of algae. Mechanical crushing is often used with chemicals. Extraction cost vary for microalgae but are likely to be around $4 a pound compared with about $1 a pound for palm oil. Oil extraction methods include the following.

- **Chemical solvents:** Benzene and ether may be used or hexane extraction, which is widely used in the food industry and is relatively inexpensive. Solvents have the disadvantage associated with working with the chemicals. Exposure to vapors or direct contact with the skin may cause serious damage. Benzene is classified as a carcinogen and chemicals are flammable.
- **Soxhlet** extraction uses chemical solvents through repeated washing or percolation with organic solvents such as hexane or petroleum ether.
- **Enzymatic** extraction uses enzymes to degrade the cell walls with water which acts as the solvent and makes fractionation of the oil easier.
- **Expeller press** pushes the oil out of dried algal biomass. Food manufacturers use a combination of mechanical press and chemical solvents to extract vegetable oil.
- **Osmotic shock**, a sudden reduction in osmotic pressure, can cause cells in a solution to rupture. The oil can then be skimmed off the top.
- **Ultrasonic extraction** can accelerate extraction processes by creating cavitation bubbles in a solvent. When these bubbles collapse near the cell walls, it creates shock waves that cause the cells walls to break and release their contents.

*Component extraction*

Removing algae components from the algal biomass presents a set of possible alternatives. Usually, the highest value added products is extracted first. If the production system, including biogenetic

## Production Challenges

constrained selection optimized pigments, then the pigments would be extracted first. For example, the pink orange pigment is a carotenoid and is used in salmon feed to create the pink salmon meat.

Pigment extraction techniques purposely disrupt cell integrity, thereby removing pigment molecules from intrinsic membrane proteins. Freezing the tissue with liquid nitrogen and grinding the still frozen tissue with a mortar and pestle or blender overcomes some of the problems of working with material that produces large amounts of viscous (sticky) polysaccharides. Freeze-thawing tissue also breaks down cellular membranes but may liberate more polysaccharides. Finely ground tissue can be homogenized in organic solvent to further disrupt cellular membranes and to liberate pigment molecules.

### *Quality control*

Measurements of process quality vary with the goal for the system. Optimizing food production requires substantial monitoring, testing and assurance that the process meets FDA and sometimes organic food standards.

Automated production systems enable quality control checks continuously for all of the critical variables including especially contamination. Quality control may include monitoring for:

- **Biomass** density, color, size, structure and vitality
- **Water** temperature, Ph (acidity), dissolved $O_2$ and $CO_2$
- **Water** quality and dissolved salts and possibly metals
- **Mixing** velocity
- **Nutrient** availability for all important nutrients

Measurement of various component parameters occurs through the process of harvest, oil extraction and component separation.

### *Marketing*

Successful marketing for algal food solutions must consider:

1. Commercial algae production to save humanity is an old dream where extraordinary promises have crashed and burned not once but several times.

2. Science fiction has positioning of algae as first savior and then Frankenfood which adds to public skepticism.
3. The industry language that reinforces consumers' fear.
4. The gag factor from the social attribution of algae as icky, slicky, green slime.
5. The failure of scaled up production means marketers have had no product to offer consumers or to test tastes, textures and aromas.

Possibly the most serious marketing issue is the dismal failure of past initiatives to produce algal foods.

**Save humanity.** Over the past century, excitement for algae as a global food solution has bubbled up several times and each time has burst in ignoble fashion. In the 1890s, experts worried about Thomas Malthus' prediction that population growth would outstrip food and recommended nontraditional food sources including yeast, plankton and algae. A similar initiative came and went after World War I.

Scientists continued their search for sustainable food sources. After the Second World War, over half the world's population was impoverished and hungry and experts recommended non-conventional agriculture as a way out of the Malthusian trap. Algae emerged as the best available antidote and numerous pilot projects attempted algal production.

Researchers announced they were able to grow nutritious algae using inexpensive materials under controlled laboratory conditions in 1948. When grown in optimal conditions – sunny, warm, shallow ponds fed by simple $CO_2$ – Chlorella converted around 20% of solar energy into a plant containing over 50% protein when dried. Unlike most plants, Chlorella's protein was complete with the 10 amino acids then considered essential and it was packed with calories, fat and vitamins.[158]

The press became ebullient about algae's potential and Colliers' Magazine sketched a farm of the future where fat coils of glass pipe produced thousands of tons of protein in automated farms.[159]

Production Challenges

Experts, not to be outdone by journalists, created plausible scenarios where algae would solve world food supplies with near zero cost.

Unfortunately, researchers rediscovered Murphy's Law repeatedly and everything that could go wrong did. Instead of being robust, Chlorella turned out to be a very temperamental organism and simply stopped growing with small changes in temperature, density, light, pH and nutrients. The plant was so fragile that harvest with centrifuges damaged the biomass as did the heat necessary for demoisturizing. Chlorella's hard cell walls made it indigestible which added the cost and energy of heat or additional mechanical processing.

While most researchers gave up on their quest to solve world hunger with algae, NASA investigated the use of algae in the 1950s as a way to feed astronauts during long spaceflights. In what has been called the "Algae Race," Soviet and American projects competed to develop a self-contained aerospace life-support system that would use algae to convert astronauts' waste into clean air, water and perhaps food.[160] However, scientists were unable to solve the contamination problems and the program was scrapped.

As part of this effort, at least one research paper was published in 1961 in the *Journal of Nutrition* titled "Algae Feeding in Humans." It sums up the sparse research on algae as a human food. The U.S. Army research team examined Chlorella from Japan that was grown in ponds, harvested, centrifuged, washed, heated and vacuum dried to a green powder. Their analysis showed the composition to be: protein, 59%, fat (oils), 19%, carbohydrates, 13%, moisture, 3% and ash 6%.

The authors found that algal food supplements of up to 100 grams per day were tolerated by their five human subjects. The green algae used, Chlorella, gave a strong spinach-like flavor to the food supplemented. The most acceptable preparations were cookies, chocolate cake, gingerbread and cold milk.

Larger supplements created stomach problems but symptoms disappeared after the supplements were discontinued. The team concluded that dried algae can be tolerated as a food supplement but further processing would be necessary before it could become a major food source. These findings relegated algae to a small sector of the foods market health foods. American research on algae as a food source practically evaporated.

Fortunately for mankind, the Green Revolution began in the 1950s and foods flourished due to three nearly equally contributing factors:

1. The invention of stronger pumps for irrigation
2. New technologies for making synthetic fertilizers
3. Advances in molecular genetics which created high-yield seeds

Stronger pumps and bigger pipes multiplied exponentially and enabled farmers to heavily over-draft groundwater of irrigation. Farmers also piled on more fertilizers, pesticides and herbicides on their fields. Green Revolution wave 1 had begun and production of food grains skyrocketed.

Non-agricultural sources of food were unnecessary due to advances in food grain production. Consumers also were conditioned by science fiction, journalists and movies to distrust non-traditional food sources.

**Science Fiction.** Warren Belasco in *Meals to Come: A History of the Future of Food* chronicled the rich science fiction literature associated with food for the last century. He labels the genre "Synthetic Arcadias" and explores utopia's evil twin, the dystopian question: "Can we invent a better, indeed a perfect, world?" Dystopians distrust the convergence between mechanical and social engineering due to the law of unintended consequences.[161] The science fiction literature provides a banquet of bizarre and terrible consequences from non-traditional food sources.

Science fiction authors both popularized the concept of synthetic foods and anticipated unfavorable consumer reactions and unintended consequences such as the Killer Tomato and Frankenfoods. H.G. Wells' The Time Machine, 1895, *War of the Worlds*, 1898, and *The Food of the Gods*, 1905, Aldus Huxley's *Brave New World*, 1932 and Ward Moore's *Greener than You Think*, 1947, all warned against biotechnological panaceas.

Production Challenges

Harry Harrison's *Make Room! Make Room!* in 1966 and Paul Ehrlich's *Population Bomb*, in 1968 explicated the horrific outcomes of unrestricted population growth. Harrison's apocalyptic scenario included plankton, yeast and algae as base foods for the starving masses. Chlorella had a fishy taste so marketers decided to produce an improved version they branded as Soylent Green.

This led to the 1973 film adaption of Harrison's book, *Soylent Green*, which suggests the algal biomass culture use not only human waste but recycled humans. Even with cannibalism, the invention could not feed everyone. Water and fertilizer shortages, plague, pestilence and pesticide poisonings ruined crops and polluted water. The greenhouse effect intensified, increasing flooding, violent storms and drought.[162]

A remake of *Soylent Green* would set algal research back at least another decade. While science fiction authors were spurring public fears of Frankenfoods, people were experiencing green slime first hand in their aquariums, pools and recreational waterways. The press was eager to convey the sensational perils of algae that created deadly toxins, killer red tides and dead zones which killed all living organisms.

The algae industry sparks consumer fears with their own language.

**Industry language.** The algae industry uses several terms that scare consumers such as bioreactor, photobioreactor and biosynthetic processes. The food industry has a long and painful history of language problems that scared consumers such as genetically modified organisms, GMOs, hydroponics and irradiation. Irradiation beams low doses of electrons or gamma rays to vegetables, fruits and meats to destroy deadly organisms such as E. coli and salmonella. It also kills bacteria and extends shelf life. Irradiation is harmless but consumers consistently avoid irradiated foods when given a choice due to fears that somehow it is radioactive.

The well-documented fear consumers have for radiation motivated the use of the term algaculture systems and green solar in lieu of the common industry term bioreactors. Consumers are likely to have an aversion to eating something that comes out of a bioreactor.

Bioreactors have become synonymous with wastewater remediation. Wastewater and food make an ugly pairing.

Other algal language includes photo biosynthetic processes that reflect that algae grow using sunshine and photosynthesis. Our consumer research suggests that consumers do not trust biosynthetic, probably because it sound artificial even though it is a natural process.

**Gag factor.** Instead of seeing a glass half full with tremendous potential for algae, many people tend to view it as a glass half full of icky slicky green slime. Consumers tend to associate algae with the gunk that fouls water in which they like to swim, fish and drink. Algal food marketers have to address this negative social attribution.

The gag factor threatens to stall the commercialization of algae as a food unless the common social attribution is changed. Consumers understandably display prejudice for their traditional foods and algae does not fit. Most people know algae only in its natural form which they tend to view as green slime. The idea of eating stinky slime of any color overwhelms logical arguments.

Changing consumer behavior to a new and improved food product is challenging but consumers are willing to make incremental changes such as moving from regular beef to premium Kobe beef. Consumers display substantial resistance if they perceived a big change in food type, texture, taste smell or appearance.

Consumers were unwilling to change from regular field grown tomatoes to hydroponic tomatoes for over a decade. Hydroponics seemed too foreign, too different. Similarly, the Flavr Savr™ tomato that offered both longer shelf life and more flavor received slow adoption due to its perceived GMO origins.

Hydroponic tomatoes caught consumers' imagination when marketers emphasized the vine ripening attribute as opposed to the plant growing in water. The Flavr Savr™ lost its market position to unbranded tomatoes that were similarly bred for longer shelf life but did not shout GMO in their name.

## Production Challenges

Rather than spending millions to overcome the severe handicap of ingrained adverse social attributions towards algae, the Green Algae Strategy uses the term Alnuts™ as the name of the algal biomass.[163] Alnuts sounds similar to food products with which consumers are familiar and evokes curiosity and interest rather than an aversive yicky reaction. Other consumer acceptable names will emerge.

Another promotion tactic will be making Alnuts available to top restaurants and chefs. When gourmet dishes integrate Alnuts and traditional foods, consumers will begin to warm up to the new organic biomass.

**No product.** Food companies have been severely handicapped for decades because they have had no algae-based food product to present consumers. The food industry has not been able to do consumer research because no product has been available. Similarly, food technologists had no product so they could not work on solving the taste, aroma and texture issues that are off-putting to consumers.

Fortunately, soy products have presented a surrogate model for a new food product with texturized vegetable protein that is used for imitation meats with good taste (soy is bitter in its natural state) and texture, tofu.

Algal marketing has been limited to generic ingredients and health foods. While the constituents include a wide variety of ingredients, algae have been treated as a commodity rather than a value-added product.

Marketers sell colorings, emulsifiers, pigments, picketers and many other food elements. The component products, especially medicines, pharmaceuticals and vaccines may have more value than the lipids used for biodiesel or the protein used for food. However, coproduct value will remain unknown until large-scale production begins.

# Chapter 7. Products and Pollution Solutions

One of the oldest and tiniest plants on Earth holds the potential to avoid starvation, water pollution and lack of energy for people across our planet.

If algae were a corporation, it would be a conglomerate combining Shell Oil, Florida Power & Light, Duke Energy, General Mills, Borden, Dole, Nestlé, Morton Salt, Procter & Gamble, Waste Management Systems, Southwest Water Company, Purina, Merck, Pfizer, Safeway, and Hi Health. These are the kinds of companies that will market algal products and solutions.

Most of the algal production and solutions have occurred as background noise to the food, water and energy industries because production systems have provided only trivial product. Even though algae are highly integrated in the global food system, most foods contain algae coproducts and not the oils or protein. Even if people read ingredients labels, few are aware of the meaning or source of ingredients such as agar, alginic acid or carrageenan.

For the last 50 years, algae producers have been limited to the production of a narrow range of health foods and an array of inconspicuous ingredients for food and consumer products. Today the opportunities are very different as producers examine a portfolio of potential products and pollution solutions that in combination may have many times more value than the production of a single health-food product. Populating the algae industry will be professionals in green collar careers.

*Green collar careers*

Green collar careers in the algae industry will focus on sustainable and efficient growing, production and marketing systems. Algae produce such flexible products that people with training and experience in a wide variety of fields will serve the industry.

Professionals engaged in growing and producing algal products may have a background in botany, bacteria or phycology, the scientific study of algae. However, few people trained in the U.S. have degrees in phycology because few universities have algae labs.

Consequently, algae producers may have training in cell biology, genetics, biochemistry, biotechnology, informatics, chemical or mechanical engineering, aquaculture or agriculture. Green collar careers in renewable food and energy include sustainable:

- **Food** – R&D, food technician, nutritionist, developer, designer, marketer, distribution, packing and consumer behavior;
- **Biofuels** – R&D, production, development, safety, distribution, education;
- **Animal feeds and fertilizers** – agribusiness technology, feed specialist, formulation specialist and distribution;
- **Medicines, pharmaceuticals and vaccines** – R&D, product specialist, marketing representatives, nurses and education;
- **Pollution solutions** – assessment, monitor, test, design, safety and integration.

Green collar careers will use a variety of descriptors such as specialist, technician or engineer. The professionals will be trained in sustainability, assessment, monitoring, environmental impacts and other green technologies.

Many of the tools and technologies developed in the aquaculture industry are being adapted for algal production. Reciprocally, one of the leading algal product lines is the production of algae with specific nutrients and pigment characteristics that can be grown locally or directly in the ponds where fish are raised.

Products and Pollution Solutions

Fish farmers find it relatively cheap to grow algal biomass nearby or directly in fishponds where fish feed on the protein and nutrients. Land-based fish food crops such as corn has quadrupled in price over the last few years similar to other food grains and must be transported, stored and fed daily to the fish.

Currently, most people are focused on the value proposition for growing algae as a biofuel because the resulting product has so much value, especially jet fuel, green diesel and hydrogen. An algae strain selected for growth as a biofuel feedstock may have 60% oils.

After the oil is extracted, the remaining biomass may contain 30% protein usable for food. It is also possible that the remaining biomass may contain more value than the extracted oils for components such as pigments, medicines, vaccines or nutraceuticals, Figure 7.1.

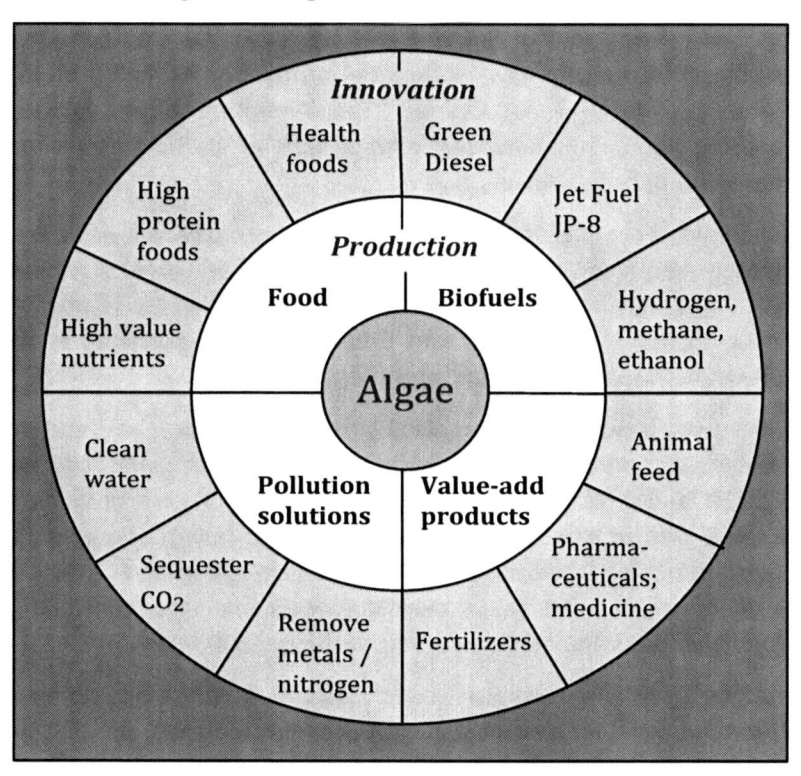

**Figure 7.1 Algae's Products and Solutions**

Inner ring: *Innovation* — *Production* — *Value-add products* — *Pollution solutions*

Outer ring segments: Health foods; Green Diesel; Jet Fuel JP-8; Hydrogen, methane, ethanol; Animal feed; Pharmaceuticals; medicine; Fertilizers; Remove metals / nitrogen; Sequester $CO_2$; Clean water; High value nutrients; High protein foods.

Center: Algae. Inner labels: Food | Biofuels.

Other algal species contain 60% protein and may be grown specifically for food such as Spirulina and nostoc are today. Indigenous peoples have used algae for food, fuel, medicines and fertilizers for thousands of years. The extraordinary food potential for algae as a protein source has been known for over 100 years.

*Food*

Humans have used algae as food and fodder since antiquity. Written records show the Chinese have harvested seaweed for more than 2,000 years and today collect over 74 species of red, green, brown and blue-green algae for food, fodder, medicines and fertilizer.[164]

Algae provide an interesting model for food because some species have over 60% protein. The four most prevalent deficiency diseases of public health importance are: malnutrition, nutritional anemia (iron and B12 deficiency), exophthalmia (vitamin A deficiency) and endemic goiter (iodine deficiency). Some digestible algae such as Spirulina address each of these issues through the production of high protein, iron and B12, vitamin A. and iodine.[165] The Kanembu tribe in Chad has been using about 10 grams per serving with most their meals for centuries with positive results.

Spirulina are more digestible that other algae because cell walls are composed of mucopolysaccharides rather than cellulose. Spirulina produce high levels of B-carotene which is converted to vitamin A during digestion. This process can prevent xerothalmia, a form of blindness that arises in malnourished children.[166]

The high nucleic acid content of algae limits their use as food because the nucleic acid produces uric acid upon digestion. Excess uric acid can precipitate and form sodium urate crystal deposits in cartilage which may cause gout or kidney stones. Human studies show that eating up to 50 grams of algae a day does not increase uric acid levels.[167] Nucleic acid levels can be lowered by careful species selection, controlling reproductive rates and heat treatment of the dried biomass.

Numerous scientists, religious leaders and marketers have extolled the food potential for a plant that produces more protein per unit of biomass at far higher speeds than the fastest growing land plant.

However, algae's high nucleic acid levels, hard cell walls, texture and taste has largely defeated attempts at using algae directly as a food.

Smart people were not wrong about algae's potential but they were ahead of the knowledge curve. Before the recent breakthroughs in biotechnology, nanotechnology, bioengineering and chemical engineering, algae's potential stayed on the horizon. Today algae's potential as a solution for world food has become realistic. World population growth, food security, water scarcity and pollution have made realizing algae's food potential mission critical.

However, until scientists engineer a solution for cell walls that makes algae digestible, algae will continue to be relegated to commodity status where it provides food ingredients rather than food eaten directly and health foods.

Consumers have not embraced algae in spite of its nutritious protein because many easily accessible, naturally occurring varieties are dried into a dark green unappetizing protein power that has a slightly fishy smell which is off-putting. Experimental attempts to modify or combine this algal biomass with other foods such as through heating, baking or mixing have not been very successful. After adding small amounts to dough for example, the consistency, color and taste become unpalatable.

Other species offer a neutral taste, various colors or no color and no smell. They may be better subjects for consumer foods research. Food technology has also improved substantially so issues such as color, consistency, texture and taste are all modifiable to some degree.

Culture represents a significant challenge for algae as a food. In developing countries where citizens desperately need protein, socio-ethnological barriers will slow adoption of algal foods. Many cultures have religious and cultural restrictions against unknown food ingredients.

Unconventional protein sources such as algae must pass a series of detailed toxicological tests before they are declared safe for human consumption. No serious anomalies were found in short-term or long-term feeding experiments or in studies on acute or chronic toxicity. All tests, including human studies, failed to reveal any evidence that would restrict the utilization of properly processed algal material.[168]

**Asian Market Seaweed Products**

Algal biomass shows promising qualities as a novel source of protein because the average quality of the typical algal species is superior compared to conventional plant proteins. Compared to land protein sources, algae are lower in fat and usually higher in fiber. Asian markets have dozens of types of seaweeds for eating, soups, stews, sushi, salads, colorings and flavorings.

Many algae are partially digestible, about 20-35%, and are often eaten in small portions. Red algae are eaten as a salad, cake or chip across the North Atlantic coast and Pacific. Called dulse, dilsk or söl, it serves as a favorite snack food in Ireland, Iceland and the Pacific Rim. It has supplied fiber and vitamins to coastal people for centuries. Dulse is a good source of minerals and vitamins compared with other vegetables and it contains all trace elements needed for humans with high protein content.[169] It can be easily harvested when the tide is out and is used for food, medicines, fertilizer and animal fodder.

**Space food.** Algae have been examined as a possible food source for bioregenerative life-support systems during long spaceflights. These systems would remediate astronaut wastes, clean the water, scrub

Products and Pollution Solutions

$CO_2$ from the air and provide $O_2$ and provide food and nutrients. Scientists developed solutions for digestible cell walls, nucleic acids and pigments (to make the food look more appealing) using standard food processing methods.[170] They were not able to solve the weight problem for the equipment required for growth and processing.

**Food production.** Many algal species support food production, in other plants, animals and fish especially aquaculture where algae feed shell and fin fish. Terrestrial algae that fix nitrogen support land plants such as rice. Biotechnologists are working on the design of pesticide resistant cyanobacteria whose nitrogen fixation processes are not repressed by environmental levels of combined nitrogen. Nostoc, for example, plays an important role in the fertilization of rice paddies by fixing atmospheric nitrogen. Herbicides and pesticides used to control pests kill the nostoc which decreases crop yields and forces farmers to apply additional fertilizer.

Algae components are commonly found in food ingredients. A normal family that uses normal dairy products may find 50% of the items in their food shopping cart contain algae ingredients. Many foods and consumer products use algae for emulsifiers and thickeners where the hard cell walls are not an issue. Valuable products come from the hard cell walls.

**Carrageenans** that make up the cell walls of several species of red and brown seaweeds are a family of linear polysaccharides. Marine seaweeds are exposed to extreme mechanical shear stress from waves and currents and developed substances that offer flexibility and toughness – carrageenans and agar. Land plants use stiff cellulose and lignin which would break in pounding surf.

**Carrageenan – Irish Moss**

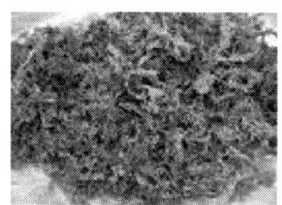

The carrageenan cell-wall material is a colloid used as a stabilizer or emulsifier and is commonly present in dairy and bakery products. Carrageenans have large, highly flexible molecules which curl forming helical structures.

After harvest, the seaweed is dried in the sun, baled and sent to a manufacturer where the biomass is ground and sifted to remove impurities. After treatment with a hot alkali solution, the cellulose is removed from the carrageenan by centrifuge and filtration. The resulting carrageenan solution is then concentrated by evaporation and is dried and ground to specification.

Carrageenan has the ability to form a variety of different gels at room temperature and is used in food procession as thickening and stabilizing agents. Carrageenan exhibit pseudoplastic properties and thin under shear stress and recover their viscosity once the stress is removed. This makes them easy to pump and they stiffen again afterwards.

There are three main commercial classes of carrageenans:

- **Kappa** — strong, rigid gels
- **Lota** — soft gels
- **Lambda** — form gels when mixed with proteins rather than water, used to thicken dairy and soy products.

Example products include: Aqua fresh Tooth Paste™ — carrageenans keep the stripe from mixing — candy bars, chocolate milk, ice cream, sour cream, puddings and pie fillings.

### Carrageenan – Red algae harvesting

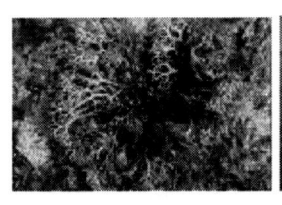

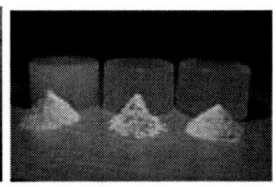

Alginates provide alginic acid from brown algae which are used to thicken liquid products and make them creamier and more stable over wide differences in temperature, acidity and time.

Kelp, Fucus and Sargassum produce alginic acid. Sargassum is fascinating seaweed that typically begins its life growing in the tropics near coral reefs. It breaks off, begins to drift and continues to grow at both ends. Ocean currents such as the Gulfstream carry it across the oceans. It is harvested from tropical waters to the British coast.

Alginic acid is extracted from the cell walls. This alginate is a colloidal product used for thickening, suspending, stabilizing, emulsifying, gel-forming or film-forming. About half of the alginate produced is used for making ice cream and other dairy products to make them smoother and prevent ice crystals. The remainder is used in other products, including shaving cream, lotions, rubber and paint.

**Alginic acid and Kelp**

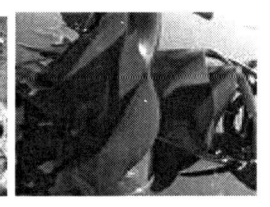

In textiles, alginates are used to thicken fiber-reactive dye pastes which facilitate sharpness in printed lines and conserves dyes. Dentists use alginates to make dental impressions of teeth. Other products include: kelp shampoo, antacids, salad dressings, syrups, orange juice and Top Ramen Noodles.

**Pigments** form a large group of algal products because algae grow a spectrum of pigments to absorb sun light. Green algae absorb all colors of light but reflects one wavelength; green. Consequently, the pigments algae uses to absorb the other colors are available in the plant.

Three major classes of photosynthetic pigments occur among the algae: chlorophylls, carotenoids (carotenes and xanthophylls) and phycobilins. Chlorophylls and carotenes are generally fat soluble molecules and can be extracted from thylakoid membranes with organic solvents such as acetone, methanol or DMSO. The phycobilins and peridinin, in contrast, are water soluble and can be extracted from algal tissues after the organic solvent extraction of chlorophyll in those tissues.

Chlorophylls are greenish pigments which algae use with chlorophyll to capture the energy of sunlight. Carotenoids are usually red, orange or yellow pigments and include the familiar compound carotene which gives carrots their color. These compounds do not dissolve in water and are attached to membranes within the cell. Carotenoids are called accessory pigments because they cannot transfer sunlight energy directly to the photosynthetic pathway but must pass their absorbed energy to chlorophyll. Fucoxanthin, also a visible accessory pigment, gives the brown color to kelps, other brown algae and the diatoms.

Phycobilins are water-soluble pigments, and are therefore found in the cytoplasm or in the stroma of the chloroplast. They occur only in cyanobacteria and rhodophyta. Green algae's pigment, beta-carotene, is used as a natural food colorant. Another natural colorant is phycocyanin, derived from Spirulina. Other pigments include lutein, zeaxanthin, astaxanthin and phycobiliproteins.

**Agar.** This substance, a polysaccharide, solidifies almost anything that is liquid. Agar is a colloidal agent used for thickening, suspending, and stabilizing. However, it is best noted for its unique ability to form thermally reversible gels at low temperatures. Agar has been used in China since the 17$^{th}$ century and is currently produced in Japan, Korea, Australia, New Zealand, and Morocco.

**Agar**

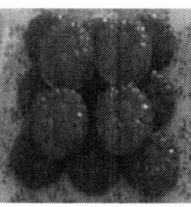

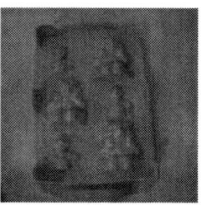

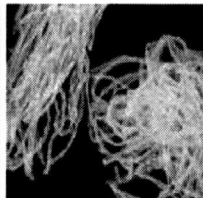

Today, the most important worldwide use of agar is as a gelatin-like medium for growing organisms in scientific and medical studies. Agar is used extensively in the pharmaceutical industry as a laxative or as an inert carrier for drug products where slow release of the drug is required. Bacteriology and mycology use agar as a stiffening agent in growth media.

Agar also is used as a stabilizer for emulsions and as a constituent of cosmetic skin preparations, ointments, and lotions. It is used in photographic film, shoe polish, dental impression molds, shaving soaps, hand lotions, and in the tanning industry.

In food, agar is used as a substitute for gelatin, as an anti-drying agent in breads and pastries and also for gelling and thickening. It is used in the manufacture of processed cheese, mayonnaise, puddings, creams, jellies and in the manufacture of frozen dairy products.

Some other products containing agar include Almost Home Cookies™, Continental Yogurt™ and Hostess™ Fruit Pies. Agar in the filling keeps the crust from getting soggy.

### Nori

Nori has been grown and cultivated around the Pacific Rim for centuries. The Japanese have developed methods for culturing and harvesting "leafy" algae called nori. In the 1970s, large farms began propagating the red alga Porphyra. Spores were sprinkled on oyster shells and placed in shallow tanks. The spores germinated and formed tiny filaments on the shells. These filaments then made their own spores, extending the filaments. Strings were dipped in the tank and the spores multiplied quickly on the strings.

**Kombu, nori and wakame**

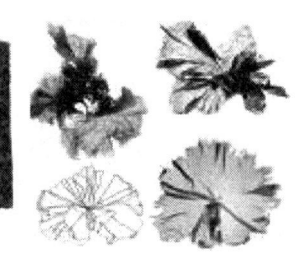

The strings were placed in shallow bays and within two months the Porphyra plants were full grown on the strings. The nori were stripped from the strings and dried in the sun. Nori sheets are eaten directly, added to soups or used to wrap rice. The popular California rolls at sushi bars are wrapped with nori.

Nori, the Japanese word for seaweed, is popular around the world but especially in Asia where it is served with a variety of names such as kombu, wakame, hai dai, laminaria and limu. Scottish cooks call it dulse and the Irish call their product dillisk.

Amanori is specifically those foods made from the Porphyra species because it contains essential amino acids, vitamins and minerals. In Korea, Porphyra, is known as kim or lavor. It provides healthy foods that are free of the sugars and fats that are associated with the Western diet.

The technique of culturing plants placed on netting horizontally was introduced in the 1960s. Netting allowed expanded production in deeper water. In the 1970s, the practice of strain selection to improve production quality became common in Japan. Growers selected plants to propagate with longer blade lengths. Ten varieties were declared in the public domain in 1980. Since then, new varieties are protected by law and users must pay royalties to the originator.[171]

### Sea Vegetables as Food

Total protein content of commercially grown Porphyra species range from 30% to 50%. The measure of value of essential amino acids is about 18, similar to whole eggs. Protein digestibility is around 76%. Some seaweed are prepared as pickled vegetables and consumed daily, similar to cucumber pickles in America. Other seaweeds are sold fresh, dried or as a flour and used in stews, soups and sushi.

**Algal Salads**

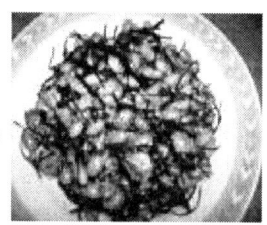

*Inland algae*

Wild populations of inland, freshwater algae have been collected and consumed since prehistoric times. One of the most common, nostoc consists of long beaded chains that form a gelatinous aggregation of filaments. The individual filaments are microscopic but aggregations occur as globules of all sizes and look similar to grapes. Some species of nostoc can fix nitrogen extracted from air. Rice paddies, for example, use the nitrogen fixed by nostoc *in-situ* by nostoc to enhance crop yields.

The microscopic filaments of Spirulina do not form oval globules but often mass into floating clumps that are pushed against the shore by wind. Other algal species appear as threads of free floating masses or filaments clinging to rocks in fast moving water.

Inland algae have not provided significant protein or energy in the diet. More commonly they are used to supplement soups, spreads and sauces and have been an important source of vitamins and minerals. They are also used to supplement animal fodder.

Spirulina, in powdered form, leads most conventional foods in both total and usable protein. Only poultry and fish are superior with more than 45% usable protein. Spirulina matches meat and dairy products with 30% to 45% protein. Spirulina and nostoc offer more protein by weight than any other vegetable.[172]

## Spirulina and Nostoc

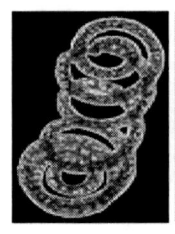

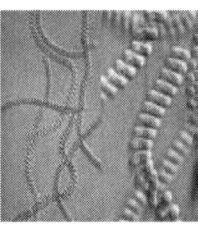

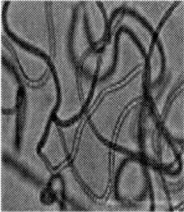

The biological value of many microalgal proteins is limited by low levels of sulfur containing amino acids. Bioengineering may provide a method for constructing algae with a full set of amino acids. Alternatively, food processing can combine foods to yield needed amino acids.

## *Kelp*

Edible kelps are a marine species of brown algae and are valued for their micronutrients especially iodine, vitamins and amino acids. Kelp carbohydrates are not digestible by humans so kelp as a food provides only roughage for the human diet.

The kelp lifecycle consists of a spore-producing stage alternating with the microscopic gamete producing stage. Even though the cells are microscopic, kelp can grow two feet a day and create considerable biomass.[173]

Laminaria, known as kombu in Japan and haidai in China, has been used for centuries in Asia. In the form of powdered kelp or kelp strips, it is used to flavor stocks for stews, soups, broths and marinades. Kelp is usually not eaten directly but gives flavor and body to the other foods. Kelp is used in rice dishes and is often served with simmered vegetables and seafood where it is eaten directly or used as a flavoring.

Many novelty kelp foods are available in health food stores and include kelp in canned salads, sweet pickled kelp and candied kelp chips. Kelp is a primary ingredient in popular snacks such as tsukudani.

Example algal food ingredients include:

- **Beer** — clarifier to remove haze-causing proteins
- **Frozen foods** – pies and pastries fillings and ice cream
- **Dairy** – whipped toppings, milkshakes, skim milk, evaporated milk, chocolate milk, ice cream, cheeses, cottage cheese, infant formulas, flans, custards, yogurt and instant breakfasts
- **High protein drinks** – protein, vitamins and minerals
- **Fruits** – fruit juices, syrups, jams and jellies
- **Sauces and gravies** – thickener
- **Pâtés** and processed **meat** — substitute fat to increase water retention and increase volume without extra calories
- **Toothpaste** — stabilizer to prevent constituents separating, e.g. striped toothpaste
- **Soy** milk – thickener emulates the consistency of whole milk.
- **Diet sodas** – clarifier

Algae have found many applications as an ingredient and in a few cases such as Nori, Porphyra and Spirulina, a positioning as a food eaten directly.

### Health foods

Algae have failed to break into the food markets because although it offers great protein values, most algae are non-digestible, look disgusting and taste dreadful. Consequently, more algal products have been sold through health food stores as pills than any other channel.

Unfortunately, too many algae consumers have bought advertising promises based on hope rather than facts. The claims made by algal health-food marketers have been largely flim-flam.

Two people were largely responsible for introducing edible algae to the Western World – British scientist, Christopher Hills and Japanese scientist Hiroshi Nakamura. Their books, *Food from Sunlight* (1978) and *Food for a Hungry World: A Pioneer's Story* (by Nakamura, 1982), outlined research efforts since 1962 and proposed edible algae as the superfood to solve the world's hunger and nutrition problems.

Unfortunately, Spirulina and other algae did not live up to its billing in the sense that research did not support many of the health claims. On the positive side, blue-green algae contain amino acids, vitamins and trace minerals that tone up the immune system, raise energy levels, and improve general health. Their high chlorophyll and phytochemical content make them effective antioxidants that help prevent cell damage and aid detoxification in the body.

**Algae Health Foods**

Beta-Carotene is known to help fight particular types of cancer and cardiovascular diseases and phycocyanin strengthens the immune system and fights cancer. No clinical studies have shown benefit from the amount of these nutrients provided in pills or powders.

Fervent testimonials of improved health from edible algae may have been a placebo effect. Improved well being could be the product of the psychological benefits of subjective delusion. Health foods have a strong self-fulfilling prophecy component to their value proposition.

Taken in the recommended doses, algae pills provide negligible nutrient value. Consumer complaints of nausea and diarrhea reinforce doubt over the digestibility of algal products.

Health food shysters have made millions from their false claims for various types of algae, especially Spirulina. Their claims included solutions for Alzheimer's disease, heart trouble, skin disturbances, allergies, prostate problems, lack of sex drive, emotional problems and alcoholism. Repeated court cases show that manufacturers and distributors hid behind their health claims by saying they were selling a food rather than a drug. Judges did not buy that argument because the food value of their pills or powders typically exceeded $300 a pound.

The FDA concluded that there is no evidence that Spirulina, phenylalanine, is effective as an appetite suppressant. The FDA has also noted that the "65% protein" claim for Spirulina is meaningless because, taken according to their label; Spirulina products provide negligible amounts of protein.

In 1982, Microalgae International Sales Corp. (MISCORP) and its founder, Christopher Hills, agreed to pay $225,000 to settle charges that they had made false claims about Spirulina. The company claimed that its Spirulina was effective for weight control and had therapeutic value against diabetes, anemia, liver disease, and ulcers.

Light Force, also founded by Hills, marketed Spirulina products with claims that they could suppress appetite, boost immunity, and increase energy. Company sales materials claimed that Spirulina is a "superfood" and "works to cleanse and detoxify the body." No evidence has been produced to support these claims.

Health food algal products contain no nutrients that are not readily available from other foods or ordinary dietary supplements that cost much less. Studies performed in countries where malnutrition is common have shown that administering Spirulina as food or tablets can correct deficiencies of the specific nutrients that Spirulina contains. However, the commercially marketed algal products have no proven value for treating obesity or other human health problems and some may contain potent toxins.

The algal product manufacturing community has not only used unethical marketing with exaggerated and false therapeutic claims but produced flawed products. Cost-cutting, unscientific and unethical practices in selecting, harvesting, and processing seaweed and fresh water algae have resulted in the marketing of some products with higher than allowed levels of impurities and toxicity.

Green Algae Strategy

For instance, in 1999, Health Canada's survey revealed that many blue-green algal products, harvested from natural lakes, contained microcystins, a toxin, above acceptable levels stipulated by Health Canada and the World Health Organization. While the health food claims have been largely puffery, the biofuel potential for algae is extraordinary.

*Fuels*

Green solar produces sustainable, liquid transportation fuels that imitate fossil fuels without fossilization. Skipping the fossilization step not only saves 200 million years of pressure and heat but lowers production costs significantly. The resulting algal oil offers many ecological advantages, especially for cleaner air.

Biofuels present an engaging value proposition for algae because various species display different lengths of hydrocarbon bonds that determine the energy value of the lipids produced. The longer length hydrocarbon bonds produce more energy. Some algal species produce lipids and 20% of biomass range while others produce over 80% lipids.

**Algal Oil**

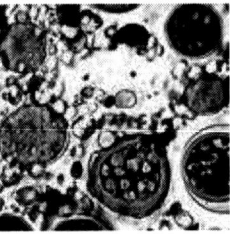

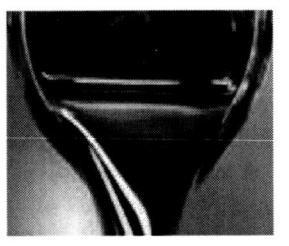

The combination of long hydrocarbon chains, explosive growth and high lipid content make algae an ideal production model for biofuels. Algal lipids can be processed to create a variety of biofuels:

- **Ethanol** – matches the current weak additive derived from corn
- **Methanol** – a light alcohol also called wood alcohol
- **Gasoline** – fits into the current model for propelling vehicles
- **Green or biodiesel** – runs directly in existing diesel cars and trucks and other vehicles and produces a near zero carbon footprint

- **Jet fuel** – enables commercial and private planes and military planes, ships, trucks and tanks to operate very demanding motors in extremely high and low temperatures
- **Hydrogen** – may provide a solution for fuel cells and high-efficiency transportation

Most companies in the algal biofuel industry are targeting the production of biodiesel because it is an environmentally friendly fuel as it reduces $CO_2$ emissions and generates less particulate matter than fossil fuels. Green diesel can fuel diesel vehicles directly with no engine conversion.

Biodiesel is made by converting bio-oil feedstock such as algal oil through a simple chemical process called transesterification. The vast majority of U.S. biodiesel today is created from soy oil. Other common feedstocks include canola, palm and waste recycled cooking oil.

The variety of the algal biofuel mix makes commercial scale algal production important for the success of continued U.S. aviation and military operations. DARPA has recognized the critical need for aviation fuels and issued a request for proposal to potential jet fuel producers for low cost production of jet fuel from algae near military bases globally.

Biofuels made from algae or other feedstocks are unlikely to be produced at a lower cost than solar, wind, waves, geothermal and other renewable sources of energy that create electricity. New solar and ocean current solutions such as solar-thermal, resonating oscillator and low-head turbine technologies will provide even lower cost electrical energy. However, the world will have the continued need for liquid fuels for surface, ocean and air transportation. Algal biofuels can provide a substantial component of this need.

After the lipids are extracted, the remaining algal biomass offers substantial value, especially for medicines, vaccines, pharmaceuticals, nutraceuticals and fertilizers.

### *Fodder – animal feed*

Dinosaurs fed on algae as did ancient sea cows, fish and whales. Indigenous people around the globe have been using algae harvested from oceans, bays, lakes and streams to supplement animal feed for centuries. Nutritional and toxicological evaluations demonstrate the suitability of some algal biomass as a valuable feed supplement or substitute for conventional protein sources such as wheat, corn, soybeans, fish meal, rice bran or hay.

Three target animal feeds are poultry, dairy and beef. Adding algae to poultry rations offers a high-value substitute for traditional grains. Beef and dairy production has the advantage that higher proportions of algae can be added to cow diets because ruminants can tolerate more as their four stomachs digest the cell walls.

Millions of acres of forests and grasslands are currently being destroyed by grazing animals, especially sheep and goats that eat plants so low to the ground they often die. Algal fodder production at the village level could save forests and grasslands by providing high quality animal feed.

### *Fish food*

Possibly 30% of current world algal production goes to aquaculture.[174] This number is probably low because many aquaculture farms now grow algae directly in their fish ponds or nearby to minimize the high cost of food grains and transportation. Growing algae *in situ* with the fish avoids the cost of buying and transporting food grains that have tripled in price. Algae provide a favorite food for fish since they feed on algae in their natural settings.

### *Fabrics*

Algal biomass includes the same carbohydrates as trees and grasses. This material can be used to fabricate textiles, paper and even building materials using glues and composites. Paper pulp, for example, is a major cause of forest destruction. Large-scale algal production could save forests by supplying pulp to make paper.

University of Texas researchers Malcolm Brown and David Nobles took blue-green algal species that did not produce cellulose and added genes from an acetobacter species that allowed them to synthesize cellulose and sugars.[175] These products could be feedstocks for textiles, paper or ethanol.

Brown and Nobles found that the cyanobacteria cellulose was a gel-type instead of the crystalline form found in the cell walls of plants and trees. Their cellulose had zero crystallinity and a low molecular weight which made it easier to break down than plant cellulose.

Cyanobacteria make a good production choice because they grow more rapidly than green algae, doubling in just four hours compared with 24 hours. Another advantage of using cyanobacteria is that some species secrete the cellulose and sugar into their surroundings where it can be harvested without sacrificing (harvesting) the organisms. Green algae has to be collected and their cells broken open to get at the oil inside. This eliminates the two most costly steps in the production.

Algal oils can be refined into any products produced by oil refineries, including plastics that are biodegradable. It may be possible to use biodegradable algal plastic to construct algaculture systems.

*Value added products*

Possibly the highest value algal coproducts are medicines, pharmaceuticals, vaccines and nutraceuticals. Algae are used to make capsules and tablets, stabilizing agents for stomachs, suppositories, radiology agents, anticoagulants and antiulcer lotions and creams.

Published research on algal medical applications include: infections, external wounds, pesticide poisoning, obesity, cancer, diabetes, hepatitis, pancreatitis, cataracts, constipation and allergies.

A large number of antibiotic compounds, many with novel structures, have been isolated and characterized in algae. Cyanobacteria have been able to produce antiviral, antineoplastic and other pharmacologically active compounds.[176] A variety of algal compounds are active against herpes virus, pneumonia virus and HIV. Some produce antitumor and antifungal compounds.

Bioactive algae compounds are finding applications in both human and veterinary medicine and in agriculture. Antioxidants provide value for the food industry. Lipophilic antioxidants serve as food preservatives, preventing lipid peroxidation which causes food spoilage. Other applications include research tools or structural models for the development of new drugs.

Algae are attractive as natural sources of bioactive molecules since they have the potential to produce compounds in a culture which enables the production of structurally complex molecules which are difficult or impossible to produce by chemical synthesis.

Single-celled eukaryotic algae have served as the model organism for the current understanding of chloroplast function, mechanisms of gene regulation and assembly and function of the photosynthetic mechanisms. The generation of transgenic algae for the expression of recombinant proteins, especially therapeutic proteins, has several advantages over existing methods. Chloroplasts can fold and assemble complex mammalian proteins quickly.

Therapeutic and diagnostic pharmaceuticals based on algae recombinant proteins have been developed recently, including vaccines, antibodies, hormones and enzymes. Strong anticoagulants also have been extracted from red seaweed.[177] Harvard Medical School reported that their algal product had restored light vision to blind mice. Light vision is the first step towards full vision recovery.

Scientists have found algal species that contain compounds that inhibit the division of cancer cells grown in the laboratory. Additionally, a compound isolated from algae collected from oil platforms in the Gulf of Mexico has been shown to block cell division and enhance the activity of the cancer drug Taxol.[178]

Intensive R&D is focused on technologies that will enable algae to serve as medical protein factories. Recombinant human proteins expressed in plants tend to show the same activity as the original protein. Algal cultivation is faster, e.g. days versus months, and less costly than using higher plants or animals as biofactories. One of the major difficulties with using plant models is that the recombinant protein must be purified from the plant material and those activities takes time and energy. Proteins are easier to separate from single-celled organisms such as algae.

Besides the production of high-value proteins for pharmaceuticals, algal chloroplasts have the potential to grow novel metabolites and to accumulate novel compounds with potentially beneficial nutritional value. Success in engineering new metabolic pathways into higher plants' chloroplasts suggests the potential of chloroplast engineering in algae.[179] This would allow algae to mimic the vitamin production pathways used by higher plants. New techniques in genetic manipulation will add new tools for the genetic engineering of the chloroplast.[180]

Algal chloroplasts appear to be an ideal platform for protein production. Engineering recombinant enzymes or pathways that could change metabolic profiles in chloroplasts would enable the ability to create novel proteins. These proteins might be fusion proteins, in which domains from two different proteins are joined together to make a hybrid protein of novel function. This makes possible the use of transgenic chloroplasts of green algae as a mucosal vaccine source.[181]

The cultivation of transgenic plants poses the danger of propagation of the transgenic spores or fragments. This could lead to uncontrolled expression of pharmaceutical proteins in non-target algae as well as propagation of resistance marker genes. Since algae do not produce pollen, there is no potential for the introduction of transgenes into other food crops. Cultivation in closed systems prevents the release of genetically modified organisms into the environment and allows for safe production.

Green algae fall into the category of Generally Recognized as Safe (GRAS), meaning they are safe to eat and offer a source for the delivery of therapeutic proteins.

A novel cultivation strategy uses algal strains with defects in metabolism for protein production without the need for resistance marker genes. Strains of algae with defective cell walls prevent unintentional propagation of transgenic material because the cells do not survive outside the carefully controlled environment. The cell wall defective proteins enable the recombinant proteins to be targeted with precision. Another advantage of the defective cell wall algal model is that the target protein can be separated from the surrounding medium faster and easier than other systems.

**Fertilizers.** Algae produce a biomass that, after coproduct extraction, can be used as a natural fertilizer and that is very low in energy production cost. Seaweed has been used as a fertilizer and soil conditioner for centuries. Currently, fertilizers consume huge amounts of energy, typically natural gas and pressure, to extract nitrogen from the air. Nitrogen-fixing algae extract nitrogen from the water and air with only the energy used to mix the algae so that it can get sunshine. Consumers like algae fertilizers because they are all natural.

Maerl is commonly used as a soil conditioner throughout Britain, Ireland and France. It is dredged from the sea floor and crushed to form a powder. The product carries considerable calcium and magnesium carbonate and works well as an organic fertilizer.

**Maerl, Fertilizer Distribution and Lawn Fertilizer**

Agroplasma is the distributor of one of Europe's leading algae-based liquid plant nutrition supplements. The company claims Ferticell is a product made of 16 unicellular bacteria types that penetrate into the leaf instantaneously. It stabilizes the auxins and cytokinins in plants which is important for plant growth and for homogeneous crop production.

Products and Pollution Solutions

**Diatomaceous Earth.** This product comes from large fossil deposits of planktonic algae called diatoms. One of the largest sites of diatomaceous Earth is in Lompoc, California. This material is actually the silica cell walls of these protists. The cell walls have minute pores and are used as an abrasive or filtering agent. Products containing diatomaceous Earth include the hygroscopic, water absorbing, packaging that comes with camera lenses and other products to moderate the effects of humidity, silver polish, rubbing compounds and shoe polish. Diatomaceous Earth is used in industrial filters and is commonly used in home pool filters.

Limestones, dolimites and limes are used worldwide in many industrial processes and applications. Limestone and dolomite are used for building stone, sculptural material and several types of cement. Polarized limestone is used as a nutritional additive in animal feed and some flours for human consumption. Calcium carbonate provides many uses including scrubbing the acidity from industrial plant stack effluents.

Other algal products include:

- **Adhesives** – glues and epoxies
- **Air fresheners** – gels and cleaners
- **Biotechnology** — gel to immobilize cells and enzymes
- **Castings** and impressions and conductivity bridges
- **Chromatographic media** – constituent separations
- **Fire fighting foam** — thickener to cause the foam to become sticky
- **Paper sizing and coatings** and textile printing and dyeing
- **Pest control** – fleas and other insects
- **Pharmaceuticals** — used as an inactive excipient in pills and tablets
- **Shampoo and cosmetic creams** — thickener

Besides providing a wide variety of coproducts, Algae provides novel, low-energy pollution solutions for water and air.

## Pollution solutions

Pollution solutions include water displacement and replacement. Algae displaces irrigation water by growing food or biofuel feedstocks that otherwise would consume water. For example, if algae were to replace corn as the U.S. biofuel feedstock, over 40 million acres of cropland would be saved plus two trillion gallons of irrigation water. Algae would also avoid millions of tons of $CO_2$, nitric oxides and other greenhouse gasses produced growing corn and the coal-fired power used to refine ethanol. If algae were to displace 10% of the U.S. grain exports while growing in brine or wastewater, another 20 million acres of cropland would be saved plus a trillion gallons of fresh water.

Algae may also replace polluted water with fresh water through bioremediation. Sewage treatment removes inorganic nutrients and toxins. Growing algae in human wastewater to produce natural fertilizer would remediate millions of gallons of water and save millions of cubic feet of natural gas used to make synthetic fertilizer.

Cleaning agricultural wastewater enables algae to feed on the nitrogen and potassium from the waste stream which saves production costs, yields rich biomass and clean water.

**Algae Wastewater Treatment**

Similar remediation of industrial wastewater would produce millions of gallons of algae oil and tons of fertilizer while cleaning millions of gallons of water. Various algal species can absorb and clean water of heavy metals such as cadmium, nickel, zinc, copper and lead.[182]

Industrial waste heat in water acts as a catalyst which speeds biomass production while it cools and cleans the water. Industrial manufacturing plants as well as coal-fired power plants and nuclear energy facilities use extensive fresh water for cooling. The excess heat can be flued through algal ponds which absorb the heat and turn the heat energy into strong bonds in the algal biomass.

Food grains consume about 1,000 tons of water for every ton of grain. Closed-loop algaculture systems recycle water and may use only 0.001 as much fresh water per pound of protein production as food grains. Of course, algaculture systems that use animal, industrial or human wastewater or brine water offer an even greater water savings compared to foodgrains. The water advantage alone makes a strong argument for algal food production.

Algae also offer significant potential cleaning the air with $CO_2$ sequestration. Power plants, manufacturing facilities and industrial producers that use coal or wood-fired furnaces can flue their exhaust smoke through algaculture systems. Algae use the $CO_2$ as food as they go about their business of creating high energy green biomass.

Organic carbon burial occurs when organic carbon compounds produced in the algal biomass sink without being completely decomposed. Algal organic carbon can be buried in deep anoxic ocean sediments where they may be sequestered from oxidation for thousands of years. Some algal strains are resistant to decomposition and others receive help from bacteria that make decay-resistant colloids that stay in sediments.[183]

Scientists use an algal growth potential test to assess nutrient levels of wastewater. The test operates on the premise that the maximum cell yield is proportional to the amount of the nutrient which is present and biologically available. Growth potential works especially well for nitrogen and phosphorus.

Scientists use algae as biomonitors much like canaries served coal miners due to their sensitivity to carbon monoxide poisoning. Algae are more sensitive than animals to many pollutants including detergents, textile manufacturing effluents, dyes and especially herbicides.[184] Algae provide one of the best available methods for detecting toxins from other algal species and are commonly used in specified EPA tests for water quality. Some species, with its symbiont lichen, are very sensitive to variations in air quality and are used for monitoring air pollution.

Algal assays and biomonitors provide a relative measurement of water quality and nutrient status but they are not yet capable of predicting ecological events such as algae blooms. However, these processes can detect low levels of toxins long before they are a threat to human or animal health.

## Chapter 8. Who is Producing Algae?

About 98% of the algal industry focus today is on algal biofuel production but has not produced a hundred barrels of oil. Within three years, the industry will be producing millions of gallons of algal oil and valuable coproducts. Green solar production will expand exponentially.

The green algae industry has moved from staid to super charged with breakthroughs occurring weekly.[185] Most innovations are coming from the private sector because U.S. government R&D and grants have been zero. Some of the leading players in the industry are getting $10 million in private equity but that is hardly enough to build a lab, let alone a small pilot plant. The industry needs substantial government investment to achieve Green Independence.

The industry faces two major threats:

1. **Insufficient investment** slows R&D and the world food crisis may degrade into mass migrations and war over insufficient food, clean water and energy
2. **Commercial firms control R&D** and put a lock on algal production and biotechnology breakthroughs

If private firms locked up the basic production methods, then sustainable world food solutions would exist but are likely to be beyond the financial means of the people who most need them.

Widespread adoption and diffusion require open source, public access to technologies.

Three companies – BASF of Germany, Syngenta of Switzerland and Monsanto of St. Louis – have filed applications to control nearly two-thirds of the climate-related gene families submitted to patent offices worldwide. These "climate ready" genes will help crops survive drought, flooding, saltwater incursions, high temperatures and increased ultraviolet radiation – all of which are predicted to undermine food security in coming decades.[186]

Company officials deny the climate-ready seed applications amount to an intellectual-property grab. They say GMO seeds will be crucial to solving world hunger but would not be developed without patent protections.

Monsanto, for example, makes 60% of its revenue from genetically modified seeds.[187] In 2006, over 78% of U.S. corn used for all purposes came from genetically modified seeds. Big agribusinesses are eager to control access to genetically modified algal strains.

Lack of government R&D investment has pushed the industry towards private investment that focuses predominately on biofuels. Algal solutions for food, medicines, vaccines and fertilizers are getting very little attention.

Industry activity is difficult to decipher due to extensive puffery and trade secrets. Executives talking about their companies at conferences, interviews, press releases and websites tend to make grandiose claims designed to impress investors but too often tend to be more hype and hope than fact. Neither production efficiencies nor costs are typically revealed because they are considered proprietary. The other challenge is that the industry is so new, few participants have either a background in biotechnology or a track record in producing algae.

Algae's potential has seduced many investors, including prominent Silicon Valley venture capitalist firms. De Beers Fuels, a South African company, collapsed in 2007 following several years of false promises.

## Who is Producing Algae?

Over 75% of the companies researched for this project that had high hopes in the 1980s and 1990s no longer exist.

The Department of Defense estimates that current production cost of algal oil currently exceeds $20 per gallon. A crude oil barrel contains 42 gallons. Crude oil gets changed into petroleum gas, gasoline, oils, tar and asphalt. Roughly 28 gallons of gasoline is refined from each barrel of crude oil.

Therefore, a $120 barrel of crude yields about $112 worth of gasoline when gasoline costs $4 a gallon. Algal fuel at $20 a gallon would cost $560. Obviously, algal fuel is not economic until the cost of production decreases by a factor of five.

In spite of the current cost difference, over 50 companies and 20 universities are working on algae, primarily for the production of algal oil. The organizations profiled here are sorted based on their algae growing strategy, not their investment potential. Growing strategies include open ponds, natural settings and closed algaculture systems. *BiofuelsDigest.com* tracks activity in algae and the biofuels industry.

### *Open ponds*

**LiveFuels**, based in Menlo Park, California plans to extend the Aquatic Species Program research and use open-pond algae biofactories to commercialize its technology. John Sheehan, who led the Aquatic Species Program, joined Live Fuels in 2007 as VP of Sustainable Development. Instead of attempting to convert algae directly into ethanol or biodiesel, this startup is trying to create green crude that could be fed directly through the nation's current refinery system.

LiveFuels created a national alliance of scientists led by Sandia National Laboratories, a U.S. Department of Energy National Laboratory focused on producing biocrude oil by the year 2010. The alliance is expected to sponsor dozens of labs and hundreds of scientists by the year 2010.

The company's web site displays technical exuberance in predicting that algae can produce up to 20,000 gallons of oil per acre. The company goes on to state that the entire U.S. supply of imported oil could potentially be grown on 20 million acres of marginal land.

**OriginOil**, Inc. in Los Angeles, California, received its first funding in 2005 and is developing a technology that will transform algae into a true competitor to petroleum. The company claims its patented technology will produce "new oil" from algae, through a cost-effective, high-speed manufacturing process. This supply of new oil can be used for many products such as diesel, gasoline, jet fuel, plastics and solvents without the global warming effects of petroleum.

**Seambiotic's Algal ponds**

**Seambiotic**, located in Ashkelon, Israel was founded in 2003 and produces algae for a variety of applications, including health foods, fine chemicals, medical products and biofuels.

The firm is working with Inventure Chemical and with the Israeli Electric Company, using IEC's smokestack for a source of $CO_2$ while it grows algae in eight open algal ponds.

**PetroSun** based in Scottsdale, Arizona, led by CEO Gordon LeBlanc, Jr., is publicly held. Petro Sun began their algae-to-biofuel production in Rio Honda, Texas, in April 2007. The algae farm is a network of 1,100 acres of saltwater ponds that PetroSun thinks will make 4.4 million gallons of algal oil and 110 million pounds of biomass per year. PetroSun has leased catfish ponds in Mississippi and Louisiana and will grow algae for fuel, feed and other coproducts.

PetroSun intends to extract algal oil on-site at the farm and transport it to company biodiesel refineries via barge, rail or truck. The company plans to open more farms in Alabama, Arizona, Louisiana, Mexico, Brazil, China and Australia in 2009.

## Who is Producing Algae?

The company also offers environmentally-friendly energy production: recycled energy. Their recycled-energy technologies capture the energy content of waste exhaust heat from industrial processes and internal combustion engines.

PetroSun's markets for the Waste Heat Generator include algae and other biomass facilities, oil refineries and drilling rig power plants. Roughly 60% of all fossil fuel burned for these and other industrial uses is wasted in the form of heat, pressure and polluting emissions. PetroSun plans to capture this wasted energy and convert it to usable electricity.

PetroSun also created an algae-to-jet fuel team relationship with Science Applications International. The companies are working to transition algal biofuel technology to the commercial sector with government contracts. PetroSun has made twenty acres of ponds available at its Rio Hondo, Texas facility for R&D related to an algae-to-jet fuel.

**Desert Sweet Biofuels,** led by CEO Gary Wood in Gila Bend Arizona, raise both Desert Sweet Shrimp and algae in mineral-rich fresh and brine water from under the Arizona desert. The farm raises algae to feed the shrimp and monitors growing parameters and contaminants carefully to assure healthy, tasty gourmet shrimp. The Desert Sweet Shrimp farm, applies environmental synergy as the mineral-rich effluence from the shrimp ponds is used to irrigate acres of prime olive trees, Durum wheat and alfalfa.

**Desert Sweet Shrimp Farming with Algae**

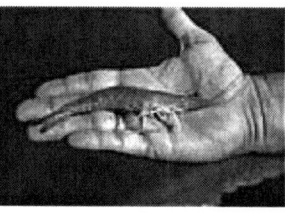

Desert Sweet Biofuels has innovated in using local agricultural wastes in a gasification chamber to produce CO for the algae ponds, $H_2$ to power an electric generator and the nutrient-rich bio-char using pyrolysis. The DSB technology is actually carbon negative because it produces the algal biomass and the bio-char that serves as an organic fertilizer while sequestering carbon.

Refiner **Neste Oil** in Helsinki, Finland is refining imported vegetable oils, palm oil and algae to make 170,000 tons of biodiesel a year in Porvoo, southern Finland. The renewable fuel is suitable for all diesel engines and is the strategic cornerstones for Neste, who say the technology outperforms both existing biodiesel products and crude oil-derived diesel products available.

Neste's renewable fuels goal is to have 70% of its raw materials coming from non-food feedstocks in ten years. By 2020, they want to have all their raw materials they use from outside the food chain.

**Ingrepo**, a Netherlands-based biotechnology company specializing in industrial large-scale algae production, plans to build algae production facilities in Malaysia. Partnering with Biomac Sdn Bhd, they will provide Malaysians with the opportunity to grow large-scale algal production for biofuels. Biomac CEO Syed Isa Syed Alwi says the algal PowerFarms will be ready for commercialization in the next year. Malaysia was chosen for its good weather conditions to grow algae, infrastructure and government interest in agro-biotechnology.

### *Natural settings*

The aquaculture industry began when producers enhanced natural settings to produce more oysters, clams and fish in open, semi- and closed-growing environments. Algae producers around the world have similarly been enhancing algal stands in natural settings by reducing predators and improving water mixing.

The advantage of finding algae growing and cultivating the growth in natural settings seems obvious: nature provides the growing container and most the nutrients. However, challenges similar to controlled settings occur in natural settings, including especially controlling growth and stability.

Who is Producing Algae?

**Kelco**, based in San Diego, harvests natural kelp beds with a specially designed mowing machine. They load the heavy biomass onto barges for transport to the processing facility to produce alginic acid.

**Neptune Industries**, based in Boca Raton, Florida, creates sustainable, eco-friendly aquaculture with integrated solutions. Dwindling supplies of wild-caught stocks, continued environmental damage, escaping fish and disease from self-polluting net pens have restricted industry growth.

In addition to hydroponically grown vegetables, lettuce, herbs and fish ponds, Neptune's patented Aqua-Sphere system uses fish waste to create additional revenue streams through the growth of algae for biofuels and methane gas.

**Blue Marble Energy,** based in Seattle, searches for unwanted wild algae growth and has developed methods for cleaning polluted water where excess nutrients lead to algal blooms that plague water systems. BME converts algal biomass to energy by creating, centralizing, and harvesting wild algae blooms.

BME's technology and process of harvesting remediates polluted water through biomass generation. BME technology harnesses nutrients and converts polluted environments into natural biofactories for generation of renewable energy feedstock while cleaning the environment. The company's business model is smart, they get paid to clean water and they produce biomass that can be processed or sold. BME marine technology can be placed in a broad array of geographies:

- Waste water systems
- Fresh water lakes, rivers, and streams
- Metal remediation for mines and other polluting industries
- Coastal remediation

By addressing wild algae growth versus the traditional mono-culture growth for biomass generation, the company keeps capital costs low and is able to produce a volume output that is multiples above closed- and pond-based systems.

**Aquaflow Binomics**, based in New Zealand, has a goal to become the first company in the world to economically produce biofuel from wild algae harvested from open-air environments. The three-year-old startup sources its algae from algae-infested polluted water systems; cleaning the polluted environment in the process.

Aquaflow Binomics harvests algae directly from the settling ponds of standard effluent management systems and other nutrient-rich water. The process can be used in many industries that produce a waste stream including the transport, dairy, meat and paper industries.

The two-step process first optimizes the ponds' productive capacity and then determines the most efficient and economic way of harvesting the pond algae. Algae are provided with full opportunity to exploit the nutrients available in the settling ponds, thereby cleaning up the water.

Algae are harvested to remove the remaining contaminants. A last stage of bioremediation, still in development, will ensure that the water discharge from the process exceeds acceptable quality standards.

The water and sludge treatment process offers a clean-up and management service for sewage treatment systems while also generating a low-cost feedstock for conversion to fuel. The result is an algae-based extract that will ultimately be converted to an alternative fuel source. Aquaflow Binomics expects to be able to produce a viable biofuel on a commercial scale.

In 2007, publicly held Aquaflow used its algae-based biodiesel to run a Land Rover driven by New Zealand's Minister of Climate Change. The company has been working with Boeing on algae-to-bio-based jet fuel.

**Biofuel Systems,** a Spanish company, is developing a system for producing energy from marine algae, with the hope of replacing fossil fuels and reducing pollution. Biofuel Systems predicts the process will produce massive amounts of biopetroleum (their term for biodiesel) from phytoplankton in a limited space and at a very moderate cost.

Who is Producing Algae?

The company says their system produces biodiesel from marine plankton and is very different from existing systems that are producing biodiesel. The company envisions producing biopetroleum using a proprietary energy converter. The system will use phytoplankton as feedstock.

## Closed systems

Closed systems offer far more control over growing parameters than open ponds or natural settings. Stressing algae to create more production of desired components by changing production parameters is practical only in closed systems. Closed systems also avoid water loss from evaporation.

**A2BE Carbon Capture** of Boulder, Colorado builds carbon capture and recycle, CCR, systems that take advantage of algae's capacity to profitably recycle industrial $CO_2$ emissions into fuel and other coproducts. Mark Allen, CEO, says their advanced energy-conversion system combines algal $CO_2$ capture technologies with biomass gasification and creates an integrated renewable fuel production system. The $CO_2$ can be recycled from any source and the biomass feedstock for gasification into syngas may come from wood waste, municipal solid waste or the processed algae waste. The $CO_2$ produced from the biomass gasification process is recycled to grow algae.

Jim Sears, President and CTO developed the patented system design and notes that the Carbon Capture and Recycle (CCR) biofactories can be scaled from a few acres to large farms that recycle industrial $CO_2$ emissions into algal biomass that can be further processed into valuable commodities including biofuel, animal feed protein and organic fertilizer.

At the core of the technology is the algae growing and harvesting biofactory. Each machine is 450' long and 50' wide consisting of twin 20' wide x 10" deep x 300' long, transparent plastic algae water-beds. It holds 150,000 gallons of algae. The biofactories work with any species of algae including cyanobacteria and diatoms.

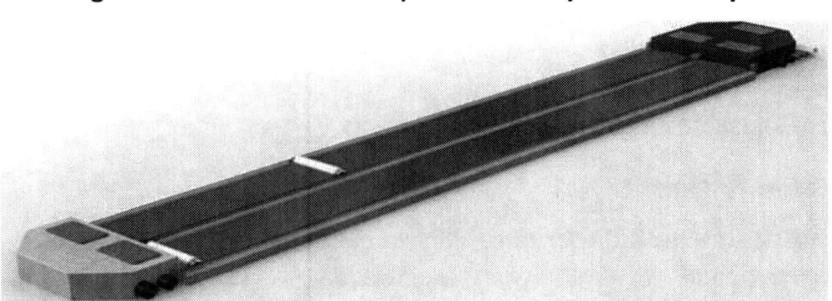

### Figure 8.1 A2BE's Carbon Capture and Recycle Biofactory

The harvesting technology is similarly adaptable to fit local needs. A2BE offers a novel bioharvesting technology where brine shrimp feed on the algae and the shrimp are harvested and processed. The CCR machine is climate adaptive due to thermal barriers above or below the culture flow that regulate temperature. This allows deployment nearly anywhere there is sunshine.

The A2BE business model shows how $CO_2$ recycling is profitable. Their business plan shows each ton of $CO_2$ may be captured at a cost of about $40 for nutrients and $10 for the CCR biofactory and operations. The net revenue of $200 per ton of $CO_2$ captured is based on: oil ($40), protein ($90), methane ($25), fertilizer ($40), oxygen ($30) and $CO_2$ credit ($25).

A2BE has created an even more compelling production attribute than the profitability per ton of $CO_2$. Their CCR biofactory creates a carbon negative process because each ton of carbon captured and recycled into the various algal coproducts displaces and avoids about 1.25 tons of carbon entering the atmosphere. The carbon negative process holds true when the original carbon is fossil sourced and the resulting products are burned as fuel.

A2BE is not only building a company to take on the substantial challenge of carbon capture but they are building a collaborative group of select institutions, corporations, and key researchers to address the spectrum of talents and disciplines needed to rapidly commercialize a solution called algae@work.

**Figure 8.2 A2BE's Vision of an Algal Farm**

**GreenFuel Technologies** of Cambridge, Massachusetts is led by Bob Metcalfe who has a telecom background. The company reached an agreement in 2008 to build a fuel plant in Europe — worth $92 M.

GreenFuel Technologies evolved from MIT and government grants for research and demonstration projects. The company has a world-class board of directors but has made some serious mistakes in executing strategy. Essentially, the company discovered, similar to many other startups, that growing algae was more expensive than they had planned. In 2007, the company had to change CEOs, lay off a large proportion of their staff and shut down some projects such as the Arizona Public Service greenhouse in Arizona.

Recent tests of an algae-based system developed by GreenFuel reported that it could capture about 80% of the $CO_2$ emitted from a power plant during the day when sunlight is available.

### Figure 8.3 Algaculture Production Systems

GreenFuel Technologies claims that using its patented technologies for growth on a one acre site the company can produce algal biomass in a year that can be separated into components that include:

- > 7,000 gallons of jet fuel
- > 5,000 gallons of ethanol
- > 1,000 tons of protein for foods
- > 200 pounds of specialized nutrients
- > 20 pounds of pigments

These production parameters lead the industry in hope and hype. GreenFuel Technologies builds algal biofactory systems which use recycled $CO_2$ to feed the algae. Their process uses the containers to carefully control the algae's intake of sunlight and nutrients. The algae are refined to biofuels. GreenFuel is backed by Polaris Ventures, Draper Fisher Jurvetson and Access Private Equity.

**Solazyme, Inc.**, based in San Francisco, is a five year old biotechnology company that harnesses the power of microalgae to produce clean and scalable high performance oils, biofuels, and "green" chemicals. The company focuses on new methods to improve production productivity.

Solazyme ignores the sun and grows algae in the dark in large tanks where they are fed sugar to supercharge their growth. Harrison Dillon, a geneticist and patent lawyer who serves as the company's president and chief technology officer claims it's a thousand times more productive than natural processes.[188] Solazyme says it has already made thousands of gallons of high-grade biodiesel and even light sweet "biocrude" with its processes, which can use anything from chemical waste to wood chips as a source of carbon.

Solazyme, raised $10 million in equity financing and $5 million in debt in 2007, and is experimenting with different feedstocks, algal species and oil extraction methods. The company hopes to reach commercial-scale biodiesel production in two or three years. Refiner Imperium Renewables of Seattle and Chevron have recently signed partnership agreements with the company.

Solazyme is using its technology to make specialty oils for the cosmetics industry in order to meet cash flow commitments. Solazyme demonstrated to the Department of Defense that their algal diesel, Soladiesel has superior cold weather properties to any commercially available biodiesel and is more suitable for cold weather climates where the military has been unable to use biodiesel.

**Algenol Biofuels** of Fort Meyers, Florida, was founded in 2006 to develop industrial-scale algaculture systems to make ethanol from algae on desert land using seawater and $CO_2$. Algenol uses a patented technology with blue green algae, cyanobacteria that are nitrogen fixing which reduces their fertilizer cost. The firm uses natural and environmental selection combined with molecular biology to produce low cost and environmentally safe biofuels.

Algenol plans to make ethanol with blue green algae that produce oil and then secrete it. They will use 3.5 million biofactories to grow the algae that are three-feet by fifty-feet and shaped like soda bottles. Most algae companies are trying to make biofuels by drying and pressing the biomass to make vegetable oil that can be processed into biodiesel. Algenol will use a process to coax individual algal cells to secrete ethanol. The fuel can be taken directly from the algal tanks while the algae continue to thrive. This process uses significantly less energy than drying and pressing the biomass for oil.

Algenol signed an $850 million deal with the Mexican company **BioFields** to grow algae for biofuel. Algenol plans to make 100 million gallons of ethanol annually in Mexico's Sonoran Desert by the end of the 2009. By the end of 2012, Algenol plans to increase production to one billion gallons. The U.S. will produce about 10 billion gallons of corn ethanol in 2008 but will consume 40 million acres of cropland, two trillion gallons of fresh water and 5 billion gallons of fossil fuel.

Algenol operates the world's largest algae library in Baltimore, Maryland to study the organism that can grow in salt or fresh water, and expanding the technique to locations beyond Mexico. The company hopes to build algae-to-ethanol farms on U.S. coasts.

## *Sapphire Energy*

Sapphire Energy, based in San Diego was launched in May of 2007 and initiated a new biofuel category called green crude production. CEO Jason Pyle says his team has built a revolutionary molecular platform that converts sunlight and $CO_2$ into renewable, carbon-neutral alternatives to conventional fossil fuels without the downsides of current biofuel efforts. The end product is not ethanol or biodiesel but biocrude, renewable 91 octane gasoline.

Sapphire's fuel products are chemically identical to molecules in crude oil, making company products entirely compatible with the current energy infrastructure — cars, refineries, and pipelines. Sapphire's scalable production facilities can grow economically because production is modular and transportable. The green crude produces fewer pollutants in the refining process and fewer harmful emissions from vehicle tailpipes.

Sapphire will not reveal the type of algae they use but it is most likely a genetically modified cyanobacteria, blue-green algae. The advantage to this form of algae is that the algae secrete the biocrude oil which rises to the top of the tank and can be skimmed. Avoiding harvesting the algae saves time, cost and may be more productive if the plants secrete enough oil.

Who is Producing Algae?

**Figure 8.4 Sapphire Energy's Biocrude**

**Inventure Chemical Technology** based in Seattle is working on their patent-pending algae-to-jet fuel product and has produced algae-based fuel in 10 gallon tests. The company plans to set up a test plant to produce up to 15 million gallons of biofuel a year. The algae used for biodiesel conversion is sourced from facilities in Israel, Arizona, and Australia. Inventure expects its technology will deliver a viable ROI for companies that use algae technology for sequestering $CO_2$.

Inventure also provides expertise in both process conversion and plant design and construction.

**Vertigro Energy**, based in San Diego, is a joint venture of eco-technology companies, Valcent Products and Global Green Solutions focused on producing vegetable oil which can be used directly as biodiesel.

Valcent's High Density Vertical Growth System maximizes algae growth in a closed loop, vertical system. In addition to biofuel, the algal oil can also be used in foods, feed stocks, pharmaceutical supplies, and beauty products.

The company says 90% by weight of the algae is captured carbon dioxide, which is sequestered by this process and contributes significantly to the reduction of greenhouse gases. Valcent has commissioned the first commercial-scale bioreactor pilot project at its test facility in El Paso, Texas. The company believes it can significantly lower costs over oil-producing crops such as palm and soybean.

**Figure 8.5 Vertigo's Vertical Growth System**

**Solena**, based in Washington D.C., uses its patented plasma technology to gasify algae and other organic substances with high energy outputs. Solena's plants produce clean, reliable electricity, using no fossil fuels and no $CO_2$ emissions.

The company is talking with Kansas power firm **Sunflower** to build a 40-megawatt power plant which will run on gasified algae. The algae would be grown in big plastic containers and fed by sunlight and sodium bicarbonate, which is a byproduct of an adjacent coal plant.

Using a plasma gasifier, Solena's technology converts all forms of biomass into a synthetic gas, syngas. The syngas is then conditioned and fed into a gas turbine to produce electricity. Solena's sequestration process recycles $CO_2$ and in the process produces biomass for a continual renewable source of fuel.

**Solix Biofuels**, based in Fort Collins, Colorado, was founded in April 2006 and backed by Colorado State University's Engine and Energy Conversion Laboratory. Solix Biofuels intends to use microalgae to create a commercially viable biofuel that will play a vital role in solving climate change and petroleum scarcity without competing with global food supply.

The company announced in 2008 that it will build its first large-scale facility at the nearby New Belgian Brewery, where $CO_2$ produced during the beer-making process will be used to feed the algae.

Who is Producing Algae?

Solix says their success comes from knowing how to select the right algal species, to create an optimal photo biological formula for each species and to build a cost-effective biofactory that can precisely deliver the formula to each individual algal cell, no matter the size of the facility or its geographical location.

**XL Renewables,** based in Phoenix, Arizona, is a 2007 start-up with a patent pending algal production system called Simgae for simple algae. The company changed its name from XL Dairy Group to XL Renewables in 2007 to emphasize its focus on creating renewable energy using dairy waste streams.

XL Renewables uses common agriculture and irrigation components to produce algae at a fraction of the cost of competing systems. The XL Super Trough uses a miniature greenhouse-type process to produce the algae in laser-leveled 18-inch deep, 1,250-foot long troughs. Mechanized equipment installs the specially designed plastic liner sheets with integrated aeration and lighting systems along the six-foot wide troughs. Depending upon need and customer demand, a plastic sheet can be installed on top of the trough to make it a closed system and increase algae production during cooler temperatures.

The XL Super Trough has no moving parts and no connection points except at the end of the troughs. The water used in the process is fortified to enhance production and is pumped through the troughs to a harvest system where the algae are extracted. The water is recycled back through the troughs.

Carbon dioxide is injected periodically and after roughly 24 hours the flow leaves the trough with a markedly greater concentration of algae than when it started. Supporting hardware components and processes involved are direct applications from the agriculture industry. Re-use of these practices avoids the need for expensive hardware and costly installation and maintenance. The Super Trough System for algae biomass production is available for $25,000 per acre.

The design is expected to provide an annual algae yield of 30 dry tons per acre annually. Capital costs are expected to be approximately $45k - $60k, a 2 - 16 times improvement over competing systems.

President Ben Cloud estimates profitable oil production costs of $0.08 - $0.12 per pound. These oil costs compare to recent market prices of feedstock oils that range from $0.25 - $0.44 per pound.

XL Renewables is developing an integrated biorefinery located in Vicksburg, Arizona, 100 miles west of Phoenix. The $260 million project integrates a modern dairy operation with a biofuels plant to produce ethanol, biodiesel, milk, animal feed and compost fertilizer. The integrated biorefinery uses the dairy manure, along with other waste streams to provide 100% of the power, heat and steam needs of the project and significantly lower production costs. The company expects to produce algal fertilizer at about $300 a ton.

XL Renewables plans to sell the XL Super Trough System and Algae Biotape globally for the economical production of algal biomass to be used as an alternative feedstock to corn for biofuels production.

**Aurora Biofuels**, developed at the University of California at Berkeley, uses genetically modified algae to efficiently create biodiesel. Aurora claims the patented technology, developed by microbial biology professor Tasios Melis, creates biodiesel fuel with yields 125 times higher and have 50% lower costs than current production methods.

**Bionavitas**, Based in Snoqualmie, Washington, says it has developed patent-pending technology for the high-volume production of algae using biofactories.

Their 2007 patent application shows their competitive advantage to be the "lighting system that includes one or more light-emitting substrates configured to light at least some of a plurality of photosynthetic organisms retained in the bioreactor."[189] This sounds like fiber-optic lights embedded in the algaculture system.

**Bodega Algae,** based in Boston, Massachusetts, is associated with MIT and was founded in 2007. Bodega Algae says it has developed a patent-pending system to grow algae in algaculture systems with light and nutrients that it says is lower cost and more efficient than the current methods. However, Bodega Algae's website is now inactive.

## Who is Producing Algae?

**Cellena,** based in Hawaii, is a joint venture created by the algae-to-biofuel startup HR Biopetroleum and Shell oil. Shell has majority share of the company, which is in the process of building a demo facility on the Kona coast of Hawaii.

Cellena announced in 2008 a new process for extracting algae oil without using chemicals, drying or an oil press. The company said that its patent-pending technique uses 26 kilowatts of power to produce 12,000 gallons of algal oil per hour with a yield of 50% from the initial algal paste.

The company also constructs and operates algae biofuels plants that use effluent gases from power plants to produce renewable fuels and to mitigate emissions of carbon.

**Canada.** Backed by oil companies and utilities, Canadian researchers have plans to develop algae farms that convert $CO_2$ from oil sands projects and coal-fired power plants into biofuels, chemicals and fertilizers. A consortium led by the Alberta Research Council has completed research that suggests the algae would thrive under northern light and temperatures with an appropriate covering for winter months.

The $20 million algal project is being funded by major Canadian energy companies, including Petro-Canada, Royal Dutch Shell, EnCana Corp. and Epcor Power, a coal-dependent Alberta-based utility.

Their research indicates that for the large industrial emitters, the system could take about 30% of their emissions. Their goal is eliminating 100 million tons of $CO_2$ emissions a year; about a third of Alberta's current production of greenhouse gas emissions. The researchers include scientists from Alberta, Saskatchewan, Manitoba and Quebec; believe they can boost the productivity of the system so that $CO_2$ can be removed at a cost of about $25 a ton.

## Health foods and nutraceuticals

The nostoc commune represents a broad set of patents for a wide variety of algae food, fuel, water, pharmaceutical and health applications.

**Nostoc commune.** Filed by Fan Lu in North Carolina and others, U.S. Patent 6,667,171 describes a process for producing Nostoc formulations using a plurality of photosynthetic microns including cyanobacteria. U.S. Patent 6,579,741 discloses a method of culturing algae capable of producing large amounts of unsaturated fatty acids and phototrophic pigments and/or polysaccharides.[190]

These patents describe methods for cultivating edible nostoc commune formulations and their use for promoting health. In addition, the invention relates to methods for promoting the health of an individual utilizing the Nostoc formulations, dietary supplements, food products and/or pharmacological compositions. This invention also provides a method for cultivating Nostoc commune comprising (a) isolating and purifying Nostoc commune; (b) culturing the Nostoc commune; and (c) conditions suitable for optimal growth of Nostoc commune. These Nostoc patents, similar to other broad patents, threaten to lock up a major algal species from public use. It will take years to determine how broadly algal patents will be enforced.

Most of the companies in this health food category harvest natural stands of algae or produce Spirulina in ponds.

**Earthrise**, based in southern California, began producing Spirulina in 1982. Today, Earthrise Nutritionals' farm is the world's largest Spirulina farm. Earthrise products are marketed in 30 countries on six continents.

**Hainan DIC Microalgae Co.** of China has a joint marketing agreement with Earthrise. The two firms produce over 800 tons of Spirulina each year in open ponds. A third facility in Thailand closed in 2006.

**Cyanotech**, based Hawaii, produces natural astaxanthin and Hawaiian Spirulina Pacifica—all natural, functional nutrients that the company claims enhance human health and nutrition. The algae is grown at its

## Who is Producing Algae?

90-acre facility in Hawaii using patented and proprietary technology and distributes them to nutritional supplement, nutraceutical, and cosmeceutical makers and marketers in more than 40 countries.

**BioEarth Spirulina**, based in Italy, and **Green Valley**, based in Germany, get their product, Spirulina Maxima from producers using artificial lakes in Mexico where the product has been grown, harvested and eaten for centuries and more recently from China. The product is sold in tablet form as a health food.

**Omega Tech,** of Boulder Colorado, markets its patented algal chicken feed rich in omega-3 fatty acids. The chicken feed, called DHA Gold for docosahexaenoic acid, the long-chain fatty acid that it contains Omega Tech president, William Barclay, claims chicken meat from animals raised on DHA Gold contain five to seven times the amount of DHA in normal commercially bred chicken.

The feed is made from schizochytrium, a tiny single-cell organism dense with DHA. When harvested and dried, the algae look like wheat flour that has a golden hue. The company grows the algae in stainless steel vats. Monsanto is analyzing the product for vitamins and to fortify food or infant formula. Infant formula, unlike human breast milk, does not contain the omega-3 fatty acids.

**Dolphin Sea Vegetable Company** in Northern Ireland, established in 1993, harvests red marine algae and markets a variety of products and food supplements directly. They also sell products to kill bacteria and viruses. DSV supplies wild red marine algae to other manufacturers and carries out research into the pharmacological benefits of seaweeds and algae for various medical, pharmaceutical and health foods.

The company is developing PhycoPLEX through clinical trials to examine its effects on immune system function and modulatory action. The clinical trials are conducted at the University of Ulster, supported by a European Union grant.

## Table 8.1 Other Firms working on Algae

| | |
|---|---|
| Algae BioFuels (PetroSun) | Global Green Solutions |
| Algaen | GreenShift |
| Arare | Green Start Products |
| Aquaflow | GS Cleantech |
| Biodiesel | Infinifuel |
| Biofuels Digest | Inventure |
| Biofuel Review | Kent Sea Tech |
| Bionavitas | Kiwikpower |
| Carbon Capture Corp | OriginOil |
| Cell Tech | PetroAlgae (XL TechGroup) |
| Diversified Energy | Plaatts |
| EnAgri | Pelletbase |
| Energy Farms | Raytheon |
| Energy Update | Renewable Energy Magazine |
| Ethanol India | Texas Clean Fuels |
| Genergetics | Simplexity |
| GreenEnergy | World Oil |

*State and university actions*

Some states are taking action to fund entrepreneurial businesses in renewable energy and biofuels.

**Scripps Institution of Oceanography**, in La Jolla, California, offers undergraduate and graduate degrees in marine biology. Scientists such as Stephen Mayfield in the Department of Cell Biology are studying genetic engineering on algae for biofuel production. Greg Mitchell teaches and researchers biological oceanography.

**Woods Hole Oceanographic Institution**, on Cape Cod, Massachusetts, offers undergraduate and graduate programs in marine biology.

Who is Producing Algae?

Scientists are working on the impact of climate warming on ice algal production in the Arctic Ocean and others are researching the causes and consequences of red tides.

**Hydrogen gas production.** Scientists at the DOE's Argonne National Laboratory in cooperation with the **University of Illinois** and **Northwestern University** are working on converting algae to hydrogen gas. They are working with algae that contain an enzyme called hydrogenase, which creates small amounts of hydrogen gas. The objective is to remove the catalyst from the hydrogenase and use it during photosynthesis.

**University of Washington,** in Seattle and Friday Harbor in the San Juan Islands, offers a marine zoology/botany program for undergraduate and graduate studies. Students may take courses in the San Juan Archipelago doing field studies of natural history, adaptations, evolution, and taxonomy of algae and herbivores.

**University of Miami.** The Rosenstiel School of Marine and Atmospheric Science offer degree programs in marine biology where students study algae, sea grasses and coral reefs among other topics.

**University of New Hampshire.** Michael Briggs and the Biodiesel Group in the Department of Physics is working on cost effective algae-based technologies for biodiesel production.

**Texas.** The Emerging Technology Fund in Texas will provide $4 million to Texas AgriLife Research and General Atomics to conduct microalgal research and development. **Minnesota** has made similar grants available for renewable energy sources.

**North Dakota.** DOE has partnered with Chevron to develop higher-oil yield strains of microalgae. The Defense Advanced Research Projects Agency, DARPA, is working on a project with Honeywell, General Electric and the University of North Dakota.

**Virginia.** Old Dominion University researchers in Virginia have successfully piloted a project to produce biodiesel feedstock by growing algae at municipal sewage treatment plants. The researchers hope that these algal production techniques could lead to reduced emissions of nitrogen, phosphorus and carbon dioxide into the air and surrounding bodies of water. The pilot project is producing up to 70,000 gallons of biodiesel per year.

**Arizona State University Polytechnic.** LARB, The Laboratory for Algal Research and Biotechnology, works on all phases of algae growth and commercial production for biofuels. The lab produces various species of algae for biofuel feedstocks and tests growing, harvesting and processing variables in the lab and at their field site.

LARB directors Professors Qiang Hu and Milton Sommerfeld supervise projects a series of projects such as producing jet fuel from algae and bioremediation of waste water from human wastes, industrial wastes and dairy waste streams. Engineers built portable algaculture systems that can test the capability of various algal species that may be used to remediate waste water at a variety of sites.

Some algae in lakes and reservoirs are capable of producing toxins that may cause fish kills and affect human health. LARB is involved in several projects where waters are monitored to check for the presence of potential toxic algae. Supported by the Salt River Project and NSF Water Quality Center, the project uses molecular fingerprinting to detect and treat toxic algae that release toxic compounds in water supplies.

**ASU BioDesign Institute** and the **Global Institute of Sustainability.** Professors Bruce Rittmann and Wim Vermaas, in life sciences have been studying cyanobacteria for the past 20 years. Currently, they are researching ways to bioengineer cyanobacteria to produce biofuel. This work sponsored by British Petroleum and Science Foundation Arizona and plans to design and create organisms that store more lipids for biodiesel.

**Greenindependence.org.** The Algae Collaboratory for Sustainable and Affordable Foods and Energy at ASU Polytechnic creates a global social network for green food and fuels. This social marketing collaboratory brings together scientists, academics, practitioners and students who share knowledge and biotechnology tools to bring algae's full promise to the world. The collaboratory operates to create

## Who is Producing Algae?

sustainable and affordable foods and fuels for all people on Earth. In addition, projects focus on remediating polluted water and air and creating valuable products such as fertilizers, medicines and vaccines.

The Collaboratory supports technology development, communication, entrepreneurship, sustainability and technology transfer for scale-up for commercial algal production. Initiatives are directed to every stage of algae selection, growth and development, harvest, processing and marketing. The Collaboratory also conducts R3D on issues associated with small-scale algaculture systems that may be sited in villages, wasteland, inner cities, roof tops and back yards.

Patents and other forms of intellectual property are possibly the strongest threat to algal production besides ethanol subsidies. Patents may prevent wide adoption of new production technologies due to the cost of paying patent holders. For example, many of the most productive corn, wheat, rice other food seeds are patented and beyond the means for many farmers.

The Collaboratory is dedicated to maximizing open source solutions available to all people on the planet. Educational and research opportunities are expanding and institutions and additional programs are posted on the collaboratory site.

# Chapter 9. Future Scenarios

The only way to help the poor is to help them to help themselves.        President Harry S. Truman

Village-scale algaculture systems will unleash the energy and ingenuity of 10 million minds globally to find stronger, faster and better ways to grow and use algae.

Several events individually or in combination will spur global algal adoption. The most significant breakthrough will be developing green solar production systems that enable people globally to begin to grow algae for products of local need. The ignition switch for the movement to engage 10 million Green Masterminds in algal production will be design, development, demonstration and diffusion of micro algaculture systems for villages, apartments, rooftops and homes.

### *Village scale algaculture systems*

Large centralized algaculture farms benefit only a relatively few people and fail to address the root causes of hunger; more equitable control over food production resources, local production and cooperative engagement. Small village-scale integrated micro algaculture systems weave a tapestry of people together to produce and process algae for food, cooking fuel, fodder, water restoration and other applications.

Village scale algaculture production systems bring hope and solutions to those with the greatest needs but the least means. Micro-loans and other mechanisms to help desperately poor farmers subsist are undermined by climate change because one drought or storm ruins not just the entire crop but forfeits the farmer's investment: seeds, fertilizer, water and labor. Many subsistence farmers cannot recover from a failed crop, especially if they used a loan for farm inputs.

Green solar systems are robust in several ways. A two week interruption in production simply means two weeks less production, not a total crop loss. If a disruption occurs due to weather, politics, health, fiesta, family event or other factor, production can resume again immediately with minimal difficulty. The investment in materials and time continue to provide benefits providing a reliable green and sustainable production system.

Villagers have no money for fertilizer, the nutrients required by algae. What they do have is an abundance of human and animal wastes although these are also a source of pollution and disease. Intestinal parasites are spread by contact with wastes. These parasites can be eliminated through sanitation and waste treatment. Wastes sanitized by solar heat can be converted to compost, biogas, clean nutrients food, fodder and fertilizer.

Microalgal production systems have been tested in India, Togo and Senegal to address problems of poor sanitation, fuel scarcity and malnutrition.[191] The design consisted of several integrated elements including community toilets, an anaerobic digester and small Spirulina cultivation ponds. The digester processed sewage and other wastes producing gas for running cooking facilities and a liquid effluent that was pasteurized in solar heaters and used to fertilize algal ponds. The algae were harvested with woven cloth and dehydrated in solar dryers. The algal biomass provided supplementary protein to villagers – primarily children and pregnant and nursing women.[192]

These pilot algaculture projects found modest success but were insufficiently robust to gain support for project expansion for a litany of reasons. They failed due to costs, $10,000 per unit, technology problems, materials and culture. Two decades later, new technologies and better materials make diffusing algal production feasible.

## *Rooftop algaculture systems*

Hundreds of millions of people live in apartments or condos with a balcony or rooftop. People in U.S. inner cities suffer from obesity, diabetes and associated maladies due to malnutrition – lack of fresh fruits and vegetables. People world-wide in cities face high food prices and poor nutrition due to insufficient vitamins and minerals.

Food marketers must spend heavily to transport and distribute food to cities which pushes up food costs, uses huge amounts of fossil fuels and adds to street and air pollution. Moving the source of food production from rural to cities creates a wide set of economic benefits and could provide valuable nutrition for millions.

People may form collectives and share gardening labor and production. The value of rooftop algaculture systems would come less from the bulk food value it provides than from the essential vitamins and minerals algal biomass contains. Micro algaculture systems would enable millions of smart people independently and cooperatively to create, innovate and refine small growing systems.

## *Increasing prices*

While continued increase in fossil fuel prices is bad for the economy and consumers, price increases will push R3D for lower cost renewable fuel solutions.

The price of premier fuels such as jet fuel has increased even more than gasoline. The aircraft industry including the manufacturers of airplanes such as Boeing and Airbus, engine producers such as GE, Pratt & Whitney and Rolls-Royce and the commercial airlines are all investigating alternative sources of jet fuel that are more reliable and less expensive. The military has experienced increasing costs and decreasing availability of military grade jet fuel, JP-8. This creates a strategic need to find reliable sources of sustainable fuel.

DARPA's current RFP asks for locally grown biofuels at a production cost of less than three dollars a gallon.[193] Current prices for jet fuel are about $6 a gallon and probably triple that for algal biofuel. The constraint that the biofuel feedstock be grown locally occurs because the military would prefer to avoid the costs and risks of long-distance transportation. The requirement for locally produced biofuel makes algae the logical solution.

DARPA is likely to invest 30 times more government funds in their biofuel procurement project than the entire U.S. government has invested in algae in the last 30 years. Lessons learned from the commercial scale algal production for military fuel will benefit the entire industry – unless they are held as intellectual property, especially trade secrets.

Food riots and threats of political destabilization will push governments and institutions to look for food solutions that are more robust, healthier and less expensive. Food producers, especially meat producers, will be eager to adopt less expensive animal fodder.

### *Medicines, vaccines and pharmaceuticals*

Currently, a wide range of medical products are sourced from algae. Most of these products are extracted from naturally growing stands.

Biotechnology will enable scientists to insert specific genes in algae that produce high-value medicines, vaccines and pharmaceuticals. Due to the speed of algae growth, these products can be produced faster and at much lower cost than traditional methods. Ideally, medical products may be grown in the massive algaculture systems associated with biofuels where medical products would be coproducts of the biofuel production.

### *Texturized vegetable protein*

The invention of digestible texturized algal protein, TAP, will ignite the use of algae in foods. TAP, Alnuts, Nostoc or other trade names may be used as a meat replacement or supplement. The extrusion technology changes the structure of the protein and yields a fibrous spongy matrix that is similar in texture to meat. TAP may be

## What is Algae's Future?

presented in a wide variety of traditional food forms such as sushi, diced chicken, turkey, tuna or red meats.

Natural soy has a distinct sour taste that is very off-putting for most consumers. Fortunately, many years of food processing have erased issues of taste, smell, color and texture. Some algae species offer almost no taste and a healthy organic smell. The product takes on any flavor with which it is combined such as green leafy vegetables, squash, potatoes, rice, chicken, seafood, pork or beef.

Several major food companies are working on technologies that weave fibers through texturized vegetable proteins in order to make steaks, chops and hamburger that taste better, have more fiber and substantially less saturated fat and cholesterol. TAP will allow meat lovers to eat a synthetic meat product that offers a positive ecological footprint. The tasty meat will taste great and be less fattening. Unlike traditional hamburgers, the TAP will provide a broad set of vitamins and minerals as well as anti-oxidants.

### *Environmentally-friendly food labeling*

Food labels today convey nothing to the consumer about the environmental resources consumed or the ecological footprint of food production, processing, storage, packaging and distribution. Even certified organic foods indicate only that they are roughly 95% free of artificial fertilizers and pesticides.

Food labeling might follow the Sustainable Food Scorecard. For example, a Green Tag 10 label for Alnuts might appear as Table 9.1.

### Table 9.1 Green Tag 10 Ecological Footprint Food Label

| Parameter | Description | Score |
|---|---|---|
| 1. Small water footprint | Minimal water per pound of product | low |
| 2. Small Earth footprint | Minimal cropland used | low |

| 3. Carbon footprint | Minimal consumption of fossil fuels | low |
| 4. Imported energy | Minimal consumption of imported energy | low |
| 5. Biodiversity | Minimal impact on biodiversity | low |
| 6. Sustainable | Minimal consumption of non-renewable inputs | low |
| 7. Minimal air pollution | No or minimal greenhouse gases | low |
| 8. Minimal chemicals | Clean – no chemicals, toxins or heavy metals | low |
| 9. Ecological footprint | Minimal erosion and fertilizer, pesticide and herbicide run-off or loss to the environment | low |
| 10. Waste management | Packaging can be recycled and does not go to land fills | low |

A new labeling system would provide customers with valuable information on the sustainability of each food product. Reporting the water footprint alone would be astonishing to many consumers who have no idea that a pound of coffee or a hamburger consumes 3000 gallons of water.

The Green Tag 10 Ecological Footprint label may be scored with a simple low, medium and high. A more sophisticated version might score on a ten point scale for each factor and provide a grand total out of 100. Red meat might score 100, poultry 70, bread 40, vegetables 20 and algae 10.

This example label favors algae-based foods for illustration. A compromise food label balanced for both land and water foods could provide valuable education to consumers about the ecological impacts and sustainability of their food choices.

**Green Fork Society**

The hypothetical Green Fork Society conveys the benefits of making sustainable consumption choices, especially for food, water and energy. Members promote the Green Tag 10 Ecological Footprint label or a similar label. The Green Tag 10 label will create predictable controversy and news while it conveys its message: "Eat smart to save our planet."

Green Fork Society members share an "Earth first" mindset. Theirs is a culture where norms minimize the total ecological footprint for food and lifestyle choices. For members, conspicuous ecologically wasteful consumption is antithetical and foolish. Green Forks support sustainability in all consumption decisions to increase consumer awareness for the ecological costs associated with buying behavior.

Members know that they cannot personally distribute food to end food riots and starvation. They also know that by making ecologically smart consumption decisions they will be preserving valuable resources for others on our planet and consumers in the next generation.

Eating an ecologically-sensitive diet has 50 to 100 times more impact on global warming and sustainable water than the sum of all actions to minimize household water or energy use.

**Sustainable communities**

Home grown food and fuels promoted by the Green Forks Society may become as widespread as Victory Gardens during World War II. Individuals and cooperatives may build algaculture systems on rooftops, balconies or backyards motivated by the desire to go green and to save money, eat healthy, nutritious foods and save the planet. Local cooperatives might operate to process the green biomass into usable food and other products.

The Green Fork Society might assist people to build hybrid solar, wind and algaculture systems that use 100% off-the-grid, renewable energy. Algal production requires small electric motors (or human labor) for mixing water, supplying nutrients, monitors and extraction. These energy needs could be supplied by hybrid systems that use the sun to produce electricity. Solar power suits algae well since solar energy is only collected during the day and algae grow only when the sun shines.

Tapping geothermal sources for warming greenhouses will be a huge breakthrough and enable year-round production in cold climates.

Sustainable, Green Fork communities may operate collectively to produce possibly 50% of the food and biofuels for their community. Communities may compete for bragging rights for self-sufficiency which means vacant lots and rooftops will sprout algaculture systems. Community members will form cooperatives that specialize in specific foods, fuels or other coproducts – especially wine and beer.

*Wine and beer*

No breakthrough will spur the explosion of algaculture more than winemaking. When the proper strains of algae are developed for wine and beer, people well beyond those with Green Forks will quickly jump on the home brew bandwagon.

Home brewing with grapes or hops is expensive and cumbersome. Typically, the grape juice feedstock must be purchased as well as the sugar, yeast and other supplies and equipment. Individuals or cooperatives who homebrew algae wine or beer will find they can grow their own feedstock. Companies that sell hot tubs and aquaculture supplies are likely to gain a huge new source of business.

Home brewers will have to come up with a better name than algae wine since it lacks cachet. Brewers will also have fun describing the wine's taste since classical features like "Earthy tones" may not be appropriate.

What is Algae's Future?

### *Premier foods*

Some species of Nostoc form small odorless, tasteless balls that appear similar to caviar. These might be processed with pigments and taste to imitate caviar. The low price point would make this form of caviar very attractive for consumers. A host of other premium foods such as truffles, saffron or foiegras would also be popular.

Alnuts is a name proposed for the food component of the algae biomass after harvest. Alnuts may become the world's food. Consumers who have an aversion to algae may find Alnuts a more palatable name for their food, especially if it is formulated to look and taste like traditional foods. Alnuts may end the need to take vitamin supplements in pill form. A huge breakthrough would occur if research shows that vitamins and mineral pills are less effectively absorbed by the body than the same vitamins and minerals ingested in the form of an attractive food.

Alnuts would combine the advantages of great taste, texture and mouth appeal with high protein, low calorie foods. Alnuts formulated as cake would provide great taste, be filling and be largely empty of calories. Alnuts could be made into texturized vegetable protein and substituted for soy or other vegetable sources.

Probably the first Alnuts-based foods will be high caloric foods such as potato chips, snacks, French fries and candy bars. These high fat foods can be delivered with equal taste appeal but save on fats and calories. Alnuts-based doughnuts could provide taste, texture and nutrition.

Gourmet algal foods will develop. Restaurants will differentiate themselves by serving gourmet algal-based foods. Popular cooking magazines will showcase algal foods on their covers and give prizes for best recipes and food innovations. Hybrid gourmet and health magazines such as *Cooking Light* will be the big winners because they can showcase phenomenal foods that are low in fat, cholesterol and calories but high on visual impact, aroma, texture and taste.

Fast food restaurants will offer healthier, tastier texturized algal protein burgers, chicken, tuna, salads and sushi. Consumers will have a choice; they can buy a conventional beef-based hamburger with substantial saturated fats for $6 or an algae-based texturized vegetable protein burger for $3 that tastes better, is bigger and provides substantial health benefits.

Consumers may prefer texturized algal protein to another new meat source they will see on the shelves at about the same time; lab-grown tissue-cultured meats. The value proposition for algal foods will include organically grown, price, taste, texture, aroma, sustainability, ecological benefits and health.

Institutional foods such as meals for schools, prisons, businesses and cafeterias will be able to provide healthier foods that taste better, offer better nutrition, have higher protein and less fat at a substantial price reduction over traditional foods.

### *Churches and community*

The Green Algae Strategy began at University Presbyterian Church when James Hershauer asked the author to speak on the status of world food during mission month. Many churches are joining the green movement and making sustainability part of their mission.

Congregations and denominations support humanitarian efforts worldwide with many missionaries in the field. Churches will be especially effective at the demonstration and diffusion of small-scale, community and village level algaculture systems for fodder, fertilizer and cooking and heating oil.

Science centers, museums and zoos can support algaculture. The mission of science centers is to engage young people in math, science and technology careers. Science centers may create display algaculture exhibits that convey the value proposition for sustainable food and green solar energy. Zoos may grow algae for fodder.

Science centers, museums and zoos offer a two-step communication flow opportunity to convey the value of sustainability. Young people who visit these exhibits would almost certainly have a natural aversion to algae. However, if they were able to see the positive characteristics and actually taste the food product, they could carry that new information to their parents.

# What is Algae's Future?

## *Solar democracy*

Jim Sears, founder of Solix Biofuels, also founded a non-profit called Solar Democracy to solve a critical challenge for the algal industry. There are over 100,000 species and a nearly infinite number of strains which means many of the best algal strains for oil, protein and other coproducts have not been identified.

Solar Democracy plans to create a low-cost micro-algae isolation, incubation, assay and digital micro-photography kit for middle and high schools. These kits would be distributed with educational materials to students via corporate sponsor relationships. Students would provide the staffing for assessing algae found locally and student teams would upload collected algae science data to publicly accessible Solar Democracy web site. Universities and commercial companies would review the data for interesting species.

Each school and team would be contributing not only to science but to assist food and energy security to people globally.

## *Carbon capture and trade*

The true cost of renewable energy sources compared with fossil fuels is masked by the lack of accounting for pollution, especially $CO_2$ loading to the atmosphere. Some estimates indicate that burning coal to create electricity creates social and environmental costs that are twice the direct cost of mining, transporting, storing and burning coal. These externalities are currently not reflected in the cost of electricity.

A method for addressing these social and environmental costs, cap and trade, puts limits on carbon pollution and creates a market for trading $CO_2$ or pollution credits. Some producers have tried to off-set their pollution by planting trees. A single coal fired power plant puts out 3.7 million tons of CO2 each year. Over 160 million trees would be needed to off-set the pollution which is impractical.

Providing financial incentives for limiting carbon pollution will provide an additional source of revenue for algal producers.

## $CO_2$ pipelines

Government incentives to build $CO_2$ pipelines from power and manufacturing plants

## Incentive prizes

The X Prize Foundation announced a $100 million fund to speed the development of clean fuels in 2008. The Foundation will provide a series of $10 million prizes for transformations in biofuels, clean aviation, fuel, energy storage, the provision of basic utilities for developing nations and other categories.

X Prize CEO, Peter Diamandis, wants to spur faster innovation and include the needs of the developing world. He says that if the U.S. had declared energy independence a goal after September 11, we would be there.[194]

## End Smoke death

Smoke death represents a significant, yet largely unknown global health issue. Smoke death continues to be largely ignored by the world community. Professional and academic research on this important topic approaches zero.

Tonight, one out of two people on Earth, 3.3 billion people, will cook their supper with biomass fuels; wood, dung, agricultural residues or coal over an open fire. These biofuels used in cooking stoves emit substantial pollutants in the black smoke, including carbon particles 2.5 microns and smaller which damage the lungs, $CO_2$, nitrogen oxides, sulfur oxides and benzene. Many people also depend on solid fuels for heating in their tiny enclosed huts.

What is Algae's Future?

**Young girl in India prepares dung fuel**

Reliance on solid fuels for indoor cooking represents one of the most severe threats to public health. In 2000, indoor air pollution was responsible for more than 1.6 million deaths and 2.7% of the global burden of disease.[195] About two thirds of smoke deaths occur in South-East Asia and sub-Saharan Africa. In developing countries, only malnutrition, unsafe sex, lack of clean water and adequate sanitation are greater health threats than indoor air pollution.

Indoor air pollution ravages rural communities and poor urban dwellers that have no other choice for cooking fuel. Smoke death occurs because mothers build cooking fires that give off sooty black smoke in their small huts. Mothers may cook five hours a day while their children sit near the cook stove. Mothers and children inhale smoke – the equivalent of smoking several packs of cigarettes a day.

Inhaled cooking smoke is one of the leading causes of disability and death globally for women and children who are the most exposed.[196] Cooking smoke causes acute respiratory infections, heart disease, pneumonia, tuberculosis, low birth weights, cataracts, muscular fatigue and multiple types of cancer.[197]

# Green Algae Strategy

**Gathering firewood, cook fire smoke and smoke hole in hut**

Globally, villages, towns and cities have denuded forests and fields for miles because firewood is considered a public resource. In too many cases, forest lands cannot replenish firewood with sufficient speed and the forest or rangeland erodes rapidly to desert. Diseases of the respiratory system from cooking fire smoke make it increasingly more difficult for women to hike extended distances to gather firewood and water for their families. [198]

Given a choice, most mothers would prefer to use an alternative source of fuel that did not require them to scavenge for twigs or dung and did not put them at risk for respiratory distress and smoke death for themselves and their families.

Algal oil burns cleanly, similar to other plant oils. Algae do not have the fibrous cellulosic structure associated with wood that burns with black smoke and carries dense soot particulates. Algal oil burns hot with a flame that may have some color from the algae but no black smoke or particulates. Locally produced algal oil could provide a solution for smoke death. The remaining biomass could provide animal feed and fertilizer.

## *Forest death*

Over half of the natural forests on Earth have been destroyed, largely for construction, food production or cooking and heating fires. Without the root systems to hold the soil and the canopy to provide nutrients, deforested areas degrade quickly with severe erosion and dramatically reduced biodiversity.

What is Algae's Future?

Deforestation destroys the habitat for animals, birds and fish and often results in the expansion of deserts. Lack of forests, especially dense tropical forests, result in net increases to $CO_2$ exacerbating global warming.

Four primary causes of deforestation include clearing forests to:

1. **Grow food** – soybeans, wheat and maize in the Americas, Asia, Africa and India
2. **Grow biofuel feedstocks** – sugarcane and palm oil in South America, Indonesia and Asia
3. **Cook food** – all continents
4. **Graze food animals** – all continents

Farmers often find that after expending considerable energy clearing forest land, the soil is not as productive as traditional cropland and it becomes nonproductive absent heavy fertilizer application after a few growing seasons. If farmers had an alternative way to grow healthy and tasty food that did not require clearing forest lands, many would embrace the alternative.

Farming biofuel feedstocks such as sugarcane and palm oil similarly come at the cost of severe deforestation. The collateral damage from deforestation, soil erosion, water pollution and dead zones will motivate biofuel feedstock farmers to other feedstocks, especially if the alternatives are more productive.

If the requirement for firewood were eliminated or modified significantly for cooking and heating, villagers could replant the forests and fields and re-forest their communities. They might even replant forests with nut or legume trees to provide extra food.

Many communities overgraze public lands with sheep and goats that eat grasses so low to the ground that the grasses die. Without grass holding the soil on the forest floor, the soil erodes. The eroded soil fails to hold moisture for the trees and forest death quickly follows. The community may migrate to another forest where the process repeats and deserts expand.

Some forests are threatened by the need for paper. Algal carbohydrate is identical to wood carbohydrate and algaculture systems could produce enough carbohydrates to reduce pressure on forests from paper pulp.

*Three step solution*

A Green Algae Strategy for solving deforestation and eliminating smoke death requires three actions:

1. **Design and develop** micro-algaculture systems that can sustain production with primarily local tools and labor for human food, vitamins, nutrients, animal fodder and cooking oil production
2. **Select or bioengineer algae** with cell walls that make lipid and protein extraction easy, cell walls that are digestible to humans and lower nucleic acid content
3. Develop simple algal **processing techniques** such as a press to extract the oil and separate the usable protein and possibly carbohydrates

Reforestation and black smoke death solutions may not occur as quickly as biofuel innovations because there is less commercial value in creating small local algal production. However, the psychological value is priceless – saving mothers and children while saving forests.

The economic value of large-scale production for biofuels may provide the necessary biotechnology breakthroughs that will allow for the recovery of forests and for millions to avoid smoke death or disability.

Algae also offer an innovative tool for disaster relief.

*Disaster relief*

Global warming and rising ocean surface temperatures assure that severe storms such as hurricanes, tornadoes, cyclones, monsoons and floods will strike croplands more often and with greater force. Rising ocean levels mean costal damage will be magnified through direct storm damage and from salt invasion from storm surge.

The cyclone that struck Myanmar in 2008 serves as a model for a food disaster because in the storm destroyed nearly everything in its path:

What is Algae's Future?

- Planted rice fields
- Seeds stored for next year's crop
- Draft animals, in this case water buffaloes
- Infrastructure including levies and irrigation ditches

Many farm families also lost their homes and all their farm and food animals. The natural disaster was intensified because the military dictatorship did not allow non-governmental organizations to enter the country or distribute food. Donor countries were refused the opportunity to off-load emergency food. This combination of events created a high probability of widespread community starvation.

The loss of this year's crop, next year's seeds, draft animals and infrastructure means that it will be several years before these farmers and their families who survive can recover.

Consider an algae scenario. Assume Myanmar farmers are shown how to build algaculture systems from mostly local materials. They could set up and use the biofactories immediately and produce algae-based animal food and cooking oil in a matter of weeks. If production were sufficiently clean, they could also produce food for the village.

Local production would overcome the political problem of food distribution. Given a choice between starvation while waiting for traditional crops or getting by on new water-based protein, many farmers would choose to grow algae. Myanmar currently produces 100 tons of algae annually in the volcanic Twin Taung lakes.[199]

The disaster relief scenario requires the same three-step solution needed for ending smoke death and deforestation. Village labor using low-tech tools such as foot pumps could provide the necessary energy to sustain the algal production system.

Small-scale algaculture systems would also solve the serious problem for the many villages that sit over brackish or saline aquifers. This non-potable water could be pumped by foot pump into an algaculture system were algae would operate to absorb the salt, leaving the water clean. The algae oils could be used for cooking fuel and heating.

Algae do not provide a full solution for malnutrition because the biomass is very low on calories. Another source would have to supply calories. However, undernourished and starving victims are susceptible to infections because hunger suppresses the immune system. Evidence suggests the phycocyanin in algae stimulates the immune system which may improve survival rates for starving people.

### DARPA success

The Defense Agency let a Request for Proposal asking for commercial scale algae production locally to a military base, e.g. Hawaii, to make jet fuel for less than $5 a gallon. The business case for locally grown aviation fuel has motivated nearly every oil company, airplane manufacturer and airline to invest in algal oil production. Success by one or more producers using the same or different production models will set off a rush for producers in many countries.

### Auto carbon capture

Cars produce about a pound of $CO_2$ per mile. When someone invents a carbon capture filter for vehicle exhaust pipes, there will be a nearly limitless supply of low-cost $CO_2$ for growing algae.

The filter would have to be removable because each 100 miles traveled would add 100 pounds to the vehicle's weight and reduce gas mileage. Green consumers would put up with the extra hassle if the process were safe, easy and fast. The filters could be aggregated at service stations and used to feed algal production or sequestered underground.

### HIV vaccine model

Science does not yet have an effective HIV AIDS vaccine. However, when a viable vaccine is developed, the challenge will be production and distribution costs. Costs are likely to be substantial because production will almost certainly require expensive use of animals or land plants. After growing the animals or plants, additional expenses are required to extract a vaccine, package and distribute the doses.

## What is Algae's Future?

Recent research suggests that algae-based proteins can inhibit the entry of the HIV virus.[200] What if a HIV vaccine were grown in a designer strain of algae? These strains may use the defective cell wall technique to assure transgenic material does not escape into the environment.

Local algal production would dramatically lower the vaccine cost. Instead of extracting the vaccine, people could eat the algae directly and let their bodies metabolize the vaccine. Local production would eliminate significant distribution and packaging costs.

The same process may work for other vaccines such as mumps, measles, malaria, polio, tuberculosis and other preventable illnesses. There are many obstacles to the vaccine scenario, including bioethics, biotechnology and socio-culture issues. However, the simplicity and cost effectiveness of an algae solution would seem to make an algae-based vaccine model happen sooner rather than later.

### *Personalized drugs*

Arizona's Biodesign Institute at Arizona State University and Translational Genomics Research Institute, TGen, are working on molecular diagnostics for specific diseases. When markers are developed for personalized diagnosis, scientist will need drugs manufactured specifically to match the genetic needs of each patient and they are likely to need the drugs quickly.

Personalized drugs and advanced compounds grown in algae may provide a cost effective solution. Such a production system could produce the designer drugs in days instead of months.

While there is a critical global need for micro-algaculture systems, large-scale systems are needed too. One example that does not yet exist may be called the kelp scenario.

### *Kelp*

The high capital cost of land-based algaculture systems plus the cost of constantly feeding the biomass $CO_2$, nitrogen and other nutrients provides financial incentives for growing marine algae such as kelp in the open ocean.

Today natural and farmed kelp beds are harvested in California, Canada, China, Japan, Korea, Thailand, Australia and New Zealand predominately for alginic acid, health foods and pet foods.

Kelp grows to 180 feet if the water is deep enough and may weigh over 700 pounds. Plants may grow two feet a day and have large leaf-like structures known as blades that originate from elongated stem-like structures. Gas-filled bladders form at the base of blades and lift the blades towards the surface and the sunshine. These features are pseudo leaves, stems and bladders which have the appearance of land plants but do not serve the same specialized functions. Kelp represents an example of parallel evolution with land plants.

Imagine a modular marine growing system that begins with a 100 acre floating net. The net would be populated with small kelp plants similar to the process used in Japan for Nori. Kelp has a small holdfast structure at the bottom of the plant that allows it to hold onto rocks or in this case the net. The net could be progressively lowered enabling the kelp to receive light as it grows toward the sun.

After a few weeks of growth when the kelp becomes established, the net would be towed to open ocean and lowered to possibly 60 feet where it would get about 10% of sunlight penetration.

The net would be anchored to the bottom, allowing the kelp to grow upward while taking nutrients from the open ocean. The result would look similar to an underwater corn field but may be harvested several times a season in sunny weather because the plant biomass grows so fast. Regular harvest occurs by mowing the kelp from above, loading it onto a barge and transporting it for processing.

Kelp concentrates iodine and absorbs iron, calcium, sodium, phosphorus, magnesium and potassium. The plant is a source of vitamins A, B1, B2, C, D, E and K plus amino acids and is very low in cholesterol and a good source of dietary fiber, pantothenic acid, zinc and copper, riboflavin, folate and manganese.

# What is Algae's Future?

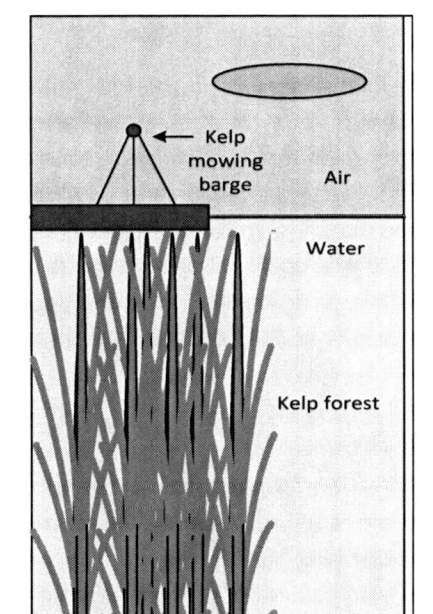

**Figure 9. 1 Kelp Farm**

Kelp's composition is similar to other algae; lipid, protein, carbohydrate and ash in a natural caloric ratio of: oils (fats), 11%; protein, 10% and carbohydrates, 79%. The biomass grows so densely, growing areas are called kelp forests in what are called trophic cascades. Kelp forests act like reefs in the ocean in the sense that they attract and protect extensive fish and other sea creatures.

The biomass production would enable the recovery of tons of carbohydrates, protein, oil and coproducts. The modular model could be replicated in various locations globally to minimize transportation costs. The kelp forest would also create a fishery that thrives in the protection of kelp. The fish also could be harvested for food.

An open-ocean kelp farm would have minimal cost. The net, harvest and anchoring are trivial costs compared with land-based algal systems. Some supplemental fertilizer may be necessary but the kelp would take most its nutrients out of the air and ocean.

This scenario would benefit if natural kelp receives some hybrid or transgenic assistance so that the caloric ratio moves closer to oils 40%, protein 30% and carbohydrates 20%. (The seaweed laver has

40% protein.) Additionally, the cell walls need to be softened to enable human and animal digestibility and it would be beneficial for the biomass to be nitrogen fixing.

Kelp would then become an excellent sustainable food, biofuel and fertilizer source. Assuming plant component modifications are successful, kelp may become the most productive food source on Earth. Transgenic assistance is difficult because a change to any plant variable creates changes in others. Even simple organisms respond in complex manners. Transgenics adds the cost of assuring the transgenic genes do not enter other organisms. Fortunately, biocontrols are easier with water-based plants that do not pollinate than with land plants.

### *Dead zones*

Dead zones occur from the nitrogen fertilizer in agricultural run-off. The rich nitrogen creates an algae bloom but the algae crowd each other out from sunshine and the underlying plants die and sink. Bacteria flourish and eat the algae and also consume all the dissolved oxygen in the water column. All life forms in the water die from oxygen starvation.

Dead zones have been doubling every ten years – largely due to increased use of fertilizers from the Green Revolution. Scientists now count 404 dead zones globally. Most countries have exploited diverse fisheries where their rivers meet the ocean but many of the fisheries have been wiped out by the dead zones.

Algaculture offers three strategies to solve the dead zone problem:

1. Algaculture or other water filtering plants such as cattails grown in low areas might clean agricultural waste streams.
2. Closed loop algaculture may displace traditional agriculture and avoid agricultural waste streams entering the water system.
3. Grow algae such as kelp in the areas where dead zones exist so the plants can absorb the surplus nitrogen and oxygenate the water. Regular harvesting would minimize the crowding problem and enable algae to flourish rather than overgrow and die.

## What is Algae's Future?

Enriching dead zones with iron to create spontaneous algae growth and oxygenation offers another possibility. As discussed earlier, scientists are mixed on spiking oceans with iron due to insufficient research on possible unintended consequences.

### *Integrated system*

The algal sweepstakes will go to the models that integrate a value chain of multiple low cost inputs with high value outputs. The input stream for algaculture is simply land, containers, sunshine, fertilizer and $CO_2$. An integrated solution might use wasteland in the desert and fertilizer from a dairy waste stream. Such an operation located near a $CO_2$ source such as a power plant, manufacturing facility or brewery, would also benefit from free $CO_2$.

Assume the algaculture production system delivers 100 tons of algal biomass per acre per year at a production cost of $1,000 a ton. This example uses common algae, not strains selected for oil or protein.

**One ton Algae – 2000 pounds**

| Component | Amount | Price | Value |
|---|---|---|---|
| Oil – 13% TAG's for biodiesel | 260 lbs | $0.60 lb | $156 |
| Protein – 77% Organic animal feed | 1540 lbs | $0.60 lb | $924 |
| Edible oils – 10% Fatty acids, omega oils, nutraceuticals | 200 lbs | $6 lb | $1200 |
| Total | | | $ 2,280 |

This model benefits from minimizing input costs by co-locating the algae farm near waste streams for algal fertilizer, nitrogen, and $CO_2$. The net $1,280 per ton means profit per acre each year is $128,000.

These coproducts may not be the best value. Experience will enable producers to reduce costs, increase production and enhance the value of the components produced.

## Market Prices for Algal Components – (2008)

| Component | Market price per lb |
|---|---|
| Medical products | $40 – $100,000 |
| Health products | $10 - $100 |
| Fortified foods, infant formula | $1 to $20 |
| Organic animal feed | $0.50 - $1.00 |
| Biofuels | $0.40 - $0.80 |
| Animal feed | $0.20 - $0.50 |
| Organic fertilizer | $0.30 - $0.50 |

The value chain elements will evolve with the industry. Successful producers will continually lower production costs, increase productivity and enhance coproducts.

The value chain elements will evolve with the industry. Successful producers will continually lower production costs, increase productivity and enhance coproducts.

The algal industry currently focuses on biofuels because the revenue models are so spectacular. Future producers may concentrate on small production of high value medical, health or gourmet food products. The future product mix for algae offers extraordinary possibilities and has yet to be defined.

# Chapter 10. What is Algae's Future?

Algaculture holds promise to provide green solutions to a rich set of Earth's most intractable challenges.

One of the tiniest and oldest plants on Earth enables algaculture and holds the potential for solutions to global hunger, fuels, fertilizers, fodder and other critical needs. The challenges facing successful algaculture probably are about as difficult as cultivating grapes and making pinot noir wine. Vintners may argue that making a great vintage is more difficult. The variables are surprisingly similar.

Today, only a few people know how to cultivate, harvest and process algae. The goal of the Green Algae Strategy is to simplify every step of the process so that anyone on Earth who wants to benefit from Algae's Green Promise can grow it easily, locally and successfully.

Commercial algal production challenges are being addressed using innovations from many industries, especially oil and gas, aquaculture, agriculture and biotechnology. The infant algal industry needs subsidies for R3D; research, development, demonstration and diffusion. Private sources are making investments but are unwilling to bear sufficient risk to move the industry forward quickly enough to meet rising U.S. and global demands for food and biofuel.

In order to execute the Green Algae Strategy, America must stop the unsustainable waste caused from pouring money into Biowar I.

## End Biowar I

The U.S. energy policy demands that taxpayers waste over $20 billion annually to subsidize accelerated global warming, the loss of U.S. fossil water and the pollution of air, soil and water. Ethanol subsidies are antithetical to American values and jeopardize world food supplies. Biowar I can end with a peace treaty that promises to withdraw not soldiers but ecologically destructive subsidies.

If corn ethanol makes sense, the market will reward it without taxpayer monies or protectionist tariffs. Closing refineries will be expensive but far less costly than continuing to throw money at an incredibly foolish act: burning food for fuel. Some refineries may stay in operation refining cellulosic feedstocks and algae.

The U.S. should look at the broader picture and withdraw subsidies that encourage ecological damage and unsustainable food production practices. The United Nations' strong voice has repeatedly recommended that countries stop subsidizing unsustainable agricultural practices.[201]

Shift those subsidies, especially for corn, irrigation water, power and Big Oil to renewable energy. End tax breaks and environmental exemptions for big polluters, especially Big Oil. Impanel a combination scientific and business group to orchestrate needed changes and enforce ecological protection regulations.[202]

## Algal production

As subsidies are shifted from ecologically destructive actions to sustainable food and biofuel production, algae will play a major role in providing solutions. The most common question people have is: "How much will it take and how fast?"

America needs to focus on renewables and make a financial commitment to R3D; research, development, demonstration and diffusion for a suite of renewable energy sources. The total commitment to renewable energy should be on the order of $50 billion annually. Quickly ending reliance on foreign fossil fuels should be priority number one for the economy and for national security.

The commitment to green solar funded at $10 billion a year, will create substantial commercial algal production within 5 years but will focus mostly on production system designs. Early years of algal production should receive, at minimum, the same financial incentives offered to ethanol producers. However, no import tariffs for biofuels are needed. The EPA need not exempt algal biofuel producers from clean air, soil and water contamination as the agency did for ethanol.

**Second wave** algal production within 10 years can displace 100% of America's imported oil with renewable and clean green solar fuel. Consumers will be reluctant to give up their gasoline vehicles but electric and diesels will become more popular with every additional uptick in gasoline prices. Trucks, trains, ships and planes will continue to consume large amounts of diesel and aviation fuel.

Second wave production will encourage individual and team entrepreneurial experimentation for home brewers who build algaculture systems for micro-scale algal production for food, biofuels and coproducts. Similar micro-scale production models will be rolling out to needy communities worldwide supported by churches and other public service organizations. Thousands of student science projects will encourage and convey the value proposition for truly sustainable locally produced foods and fuels.

Awards, special events and prices will motivate individuals, schools, universities and teams to innovate to enhance every element of algal production, harvest, processing and marketing. Open source initiatives can convey these innovations to the world.

**Third wave** production within 20 years can displace imported oil and fossil fuels used for the electrical grid and continue the important work of carbon re-capture. Sequestering 100% of fossil fuel carbon in the atmosphere and oceans may take 75 years but it is possible if algae are grown in oceans, especially in dead zones.

A huge shift will have occurred in moving transportation vehicles from liquid fuels to electric or fuel cells powered by renewable energy sources. However, America and the world will have a legacy of millions of vehicles that still need liquid transportation fuel.

Third wave production will improve the robustness and efficiencies of both macro and micro production systems. Third wave technologies will enable communities to be self-supporting for food and fuels. Barter systems may evolve where communities trade biofuel or coproducts for the luxury of traditional foods or specialty products.

Distributed production systems in communities globally will be growing food and fuels locally. Political issues will continue but the basic needs for food and cooking fuel will be provided in many communities by local green solar systems. Local production may include growing high-value vitamins, medicines and vaccines such as malaria and HIV when they are developed.

The third wave will also see the engagement of a suite of carbon neutral energy sources including wind, waves, tides, solar, geothermal and nuclear energy replacing fossil fuels used for producing electricity. These energy sources will not replace liquid transportation fuels for airplanes or ships so green solar will continue to replace fossil fuels for targeted transportation needs. The immediate future will benefit from the biotechnology associated with the recent algae genome annotation projects.[203]

### *Biotechnology*

Biofuel production with corn or land plants uses a crop that was bred for food, not energy maximization. Biotechnology may assist by providing solutions to characteristics that are intractable to breeding. These occur where exogenous genes are needed or where tissue-specific or temporal expression or suppression of endogenous genes would be valuable.[204]

Reengineering algae to express more lipids or softer cell wall material is extremely complex. Even when genes of known expression are added, the plant does not always act in a predictable manner. For example, breeding programs for Nori in China selected for thinner cell walls successfully but the plants became susceptible to fungi invasion.

Algal researchers may build on 80 years of genetics and 30 years of molecular biology to engineer efficient expressions of important therapeutic proteins. In 2004, a consortium of laboratories initiated a project to sequence the genome of Ectocarpus siliculosus, a small filamentous brown alga found in temperate, coastal environments throughout the globe. The E. siliculosus genome, which is currently being annotated, is expected to be the first completely characterized genome of a multicellular alga.[205]

Brown algae were chosen for genome analysis because at least 50 species of brown algae are used as human food and they are distributed all over the Earth. DNA analysis shows that brown algae are related to animals, fungi and green land plants. Their independent evolutionary history furnished brown algae with many novel metabolic, physiological, cellular and ecological characteristics, including a complex halogen metabolism, cell walls containing many unusual polysaccharides and high resistance to osmotic stress. Novel biomolecules such as polysaccharides and defense elicitors have made brown algae commercially valuable algae for the food industry.

The *Ectocarpus* genome project is developing several molecular tools, including mutant screens, genetic transformation and genome-scale analysis of gene expression. The availability of the genome sequence together with the ability to analyze gene function by forward and reverse genetic approaches will make it possible to address additional issues such as the biosynthesis of diverse, brown algal-specific metabolites; lipids, complex cell wall components and the genetic basis of resistance to biotic and abiotic aggression.[206]

Biotechnology will provide strong solutions but there remain a host of challenges for successful algal production. Outstanding issues present a punch list for achieving the Green Algae Strategy.

### *Challenges*

Key issues for the macro and micro production of algae are summarized in Table 10.1. The most pressing challenge lies in scaling up algaculture system size for continuous commercial production. The sparse R&D means the favored technologies have not been tested even on a pilot scale. Fortunately, much of the necessary production knowledge comes from hydroponics and aquaculture where R&D has moved those technologies forward.

## Green Algae Strategy

The challenges presented by algal production are nontrivial but commercial producers are growing algae in open ponds, estuaries of the ocean currently in California, Hawaii, South Africa, Japan, India, Myanmar, Thailand and China. Focused R&D can have scaled biofuel production systems operating in the U.S. within three years.

**Table 10.1 Priorities for Green Algae Strategy Success**

| Challenge | Issue | Resolution |
|---|---|---|
| Growth systems | Large scale, continuous production, Biofouling, Instability Micro grow systems | Growing tanks or ponds Mixing methods Light penetration Monitors Cooling and heating |
| Inputs | Water sources $CO_2$ availability / transport Nutrients, heat | Fresh and salt water Industrial wastes Wastewater treatment $CO_2$ mixing methods Nutrient sources Nutrient delivery |
| Stability | $O_2$, $CO_2$ and nutrients pH and turbidity Biomass composition | Test monitoring methods Removal methods Establish key parameters |
| Species | Selection and sources Contamination | Food or fuel? Digestibility/extractability Coproducts Productivity |
| Harvest | Filtration, sedimentation, flocculates and centrifuge | Timing Materials and methods Automation |
| Processing | Dewatering /drying Drum or freeze-drying Oil extraction options Food recovery | Extraction methods Product separation Recycle water and nutrients |

| | | |
|---|---|---|
| Oil conversion, refinement | Process efficiency – Type of oil – fatty acid profiles | Create highest value fuel Automate oil profiles Biomass composition |
| Food production | Process optimization Nutrients Digestibility | Create highest value food Automate food profiles Find more digestible strains |
| Systems integration | Wastewater, flue gases, algae culture, biomass residues and system scale up | Co-locate next to power plants or beer manufacturing plants Natural waste streams |
| Consumer acceptance | Consumers fear eating algae | Promote the social benefits of eating green |
| Social attribution | People have negative beliefs about algae | Communicate the Alnuts value proposition – sustainable food and fuel |
| Lifecycle analysis | Input/output flows Energy balance Cost benefit analysis | Examine total system Create core metrics Track feed stock market |

*Algaculture limitations*

Algae are not a complete solution to many problems and have limitations.

- **Cost.** Currently, algal production costs are about the same as red meat and most consumers today prefer meat. Comparable grains, such as soy, are produced at about 15% of the current cost of algae protein. Costs are likely to flip with large-scale biofuel production because algal protein will be a coproduct.
- **Calories.** Algae provide plenty of protein but are sparse on calories which is great for dieters but not for people suffering from malnutrition who need to find additional caloric sources. Fortunately, calories are cheaper to produce than protein.

- **Complete protein.** Algal proteins often lack some amino acids so either an additional source must be found and added to diets or biotechnology or growing conditions will have to be modified to provide complete protein. Adding additional nutrients to the growing medium may solve the amino acid deficiency.
- **Digestible cell walls and nucleic acid.** Mechanical processing, hybridization, species selection and biotechnology will find low cost, robust solutions.
- **Lack of knowledge.** Most people have little knowledge of algae other than the green slime that is too visible in pools, ponds, aquariums and waterways.
- **Negative attribution.** Most consumers carry a belief system that algae are an aversive plant with an awful smell, texture and taste. This pessimistic belief system must be overcome before consumers will embrace algae as a food source.
- **Health food claims.** Most health food claims have been false – based hope rather than facts. However, these unsupported claims occurred before biotechnology was available to design organisms that provided targeted medicines, pharmaceuticals, vaccines and nutraceuticals.
- **Energy cost.** Biofuels will not compete with cheaper sources of energy such as solar, wind, waves and geothermal. However, those renewable sources do not create liquid transportation fuels. The U.S. energy policy will apply a portfolio approach to renewable energy where biomass from algae and other feedstocks represents only a portion of the solution. The world will have a legacy of millions of vehicles running on liquid transportation fuels so biofuels will be necessary.

Possibly the major limitation to algaculture is the lack of public access to models for large- or micro-scale algal biofactories. Some of the commercial produces will be successful and will sell their technologies under patent. Micro algaculture systems represent a cornerstone of the Green Algae Strategy and will be posted for public access.

The sustainability of potential food and biofuel sources maybe measured with a scorecard.

*Sustainable scorecard*

The choice for a food and energy delivery system should have high productivity biomass with minimal resource footprint similar to the Sustainable Foods and Biofuels Scorecard that accounts for key sustainability parameters.

### Table 10.2 Sustainable Food and Biofuels Scorecard

| Parameter | Description |
|---|---|
| 1. Small water footprint | Grows with minimal water – preferably waste, grey, saline, or reuse water |
| 2. Small Earth footprint | Requires minimal land – preferably not cropland |
| 3. High energy yields | Biomass offers a high energy productivity |
| 4. High energy fuels | Convert to fuels with equal or higher energy per unit than gasoline |
| 5. Net energy yield | Gives more energy than is used to grow the biomass and refine the oil |
| 6. Sustainable | Green; continuously renewable without consumption of non-renewable inputs |
| 7. Positive air quality | Improves air quality; reduces rather than adds to GHG |
| 8. Low use of imported energy | Uses primarily natural gas and the electrical grid for biomass production and refining |
| 9. Ecological footprint | Minimal erosion and fertilizer, pesticide and herbicide run-off or loss to the environment |

| | |
|---|---|
| 10. Net food yield | After oil is removed for fuel, remaining biomass offers food value for humans and animals |
| 11. No mono-cropping | No requirement for massive production of one plant |
| 12. Economically sensible | Positive cost / benefit profile |

Plants that maximize these sustainability parameters will provide sustainable foods and fuels that will not come at the expense of traditional food crops. These sustainability parameters also mean production will take little fresh water and be a truly green product.

One of the major threats to the industry remains open source collaboration versus proprietary technologies. Should private investments enable a company to solve the cell wall issue and patent a basic technique, the world would have a great solution that was locked up in patent. Public funding needs to support basic R&D to ensure production pathways re open to people with need globally.

### Algae's value chain

Algae's potential as a food or biofuel has been largely under foot but out of mind for most people, institutions and governments. This plant has received considerable attention as a pest but few have considered its potential to serve our planet.

Algae are uniquely positioned to provide a value chain of products and solutions for critical human needs. The 16 factor value chain includes sustainable foods, fuels, ecological and novel solutions represented in **Algae's Green Promise**.

### Table 10.3 Algae's Green Promise

| Food |
|---|
| **1. Food.** Algae supply high-protein, low-fat, nutritious, healthy and delicious human foods. Algae provide more vitamins, minerals and nutrients than land plants and are a natural health food. Algae do not provide a full solution for malnutrition due to their few calories. |
| **Note:** Algae's food value will be suboptimal until solutions are found for a few key issues; making hard cell walls digestible and producing fewer nucleic acids. All other green promises await only macro and micro-scale algaculture production systems. |
| **2. Food ingredients.** Algal ingredients enhance about half the food products in a grocery store. Algae components support dairy products, beer, jams, bakery products, soups, sauces, pie fillings, cakes, frostings, colorings, ulcer remedies, digestive aids, eye drops, dental creams, skin creams and shampoos. |
| **3. Fodder.** Algae produce high-protein, low-cost, nutritious animal feed with numerous vitamins, minerals and nutrients. Replacing half the food grains fed to animals sold as U.S. exports would save 20 M acres of cropland and a trillion gallons of water. |
| Local production in villages would feed millions of animals and would save 20 M acres a year of forests and grasslands from desertification due to animal forage. |
| **4. Fisheries.** Algae provide high-protein; low-cost, nutritious fish feed, vitamins and nutrients. Algae can be grown *in-situ,* in the water with the fin fish or shell fish. Fish tend to grow with more vitality on algae than land grains because they eat algae in their natural habitat. |

## Fuel

**5. Fuels – biodiesel.** Algal oils pressed directly from algal biomass produce renewable and sustainable, high energy biofuel from sunshine, $CO_2$ and wastewater. Replacing U.S. ethanol production would take 2 M acres of desert, half of one Arizona county. Replacing corn as a biofuel feedstock would save 40 M acres of cropland, 2 trillion gallons of water, 240 M tons of soil erosion and extensive water pollution annually.

**6. Fuels – jet fuel, ethanol and hydrogen.** Algal production can be refined to a variety of high energy liquid transportation fuels including gasoline. While refining generally requires more energy input than squeezing out algal oil, the U.S. is likely to have a surplus of ethanol refinery capacity. Algal products can be refined in fossil fuel refineries into many of the products made from fossil fuels. Hydrogen gas production would occur in the algaculture system and need a refinery.

**7. Fossil fuels.** Replacing U.S. ethanol production also would save 7 B gallons of fossil fuel used to produce ethanol. Moving 1/10th of U.S. agricultural production from dirty diesel to clean algal-diesel would clean the environment and save 20 B gallons of fossil fuels annually. Even larger fossil fuel savings would accrue from using algal oils to substitute for a portion of the diesel used by trucks and trains.

**8. Fire – cooking.** Black smoke from cooking fires and heating with wood, weeds and dung causes smoke death for 1.6 M and disability for 10 M mostly women and children every year. Clean-burning, high energy algal-oil can end smoke death and the many smoke disabilities. Substituting algal oil for wood and agricultural materials will save a tremendous amount of labor from gathering firewood and allow forests to be replanted.

### Ecological Solutions

**9. Fresh water.** Running wastewater through algaculture feeds the plants and cleans the water. Producing fuel, fodder or fertilizer using wastewater or brine water saves water that would otherwise be used for land-based crops. Replacing half of U.S. food exports with algaculture foods would save 30 M acres of cropland, 4 trillion gallons of water and 15 B gallons of fossil fuel.

**10. Fresh air.** Flueing smoke stack gasses through algaculture removes $CO_2$, nitric oxides, sulfur and heavy metals such as mercury from power plant or industrial plants, sequesters greenhouse gasses and cleans the air. Algae represent only a partial solution since the plant only grows with sunshine and power plants operate 24 hours a day.

**11. Fertilizer.** Nitrogen-fixing algae may provide high nitrogen fertilizers at very low cost in both production and energy inputs. The product is natural and supports organic food production and could provide cheap local fertilizer to subsistence farmers globally. The ash retains fertilizer value after being burned in cooking fires.

**12. Forests.** High energy algal-oil fuel can end the need to denude forests and grasslands for cooking and heating fuel. Villagers may replant their forests with nut trees or legumes for food to offset the low calories provided by algal foods.

### Novel Solutions

**13. Fabrics.** Algal carbohydrates are similar to wood and may be made into textiles, paper and building materials. Algal paper and building materials save forests and fabrics and provide warmth. Algal oils may be made into biodegradable plastics or other refined products.

**14. Foreign Aid.** American foreign aid provides subsidized U.S. food, undermines or destroys local food production because farmers cannot compete with U.S. subsidized food. Gifting food fails to address the root cause of hunger and poverty – local control over food resources and community engagement. Algaculture foreign aid would transfer knowledge and some start-up materials to grow algal foods, fuels, fodder, fertilizer and medicines locally.

**15. Famine and disaster relief.** Algae, with its rich set of vitamins and minerals, may activate the immune system and ward off starvation while providing fuel, fodder, fabrics, fertilizers and fine medicines. Disaster relief with local algaculture production may prevent community starvation for millions. Local algal production solves the critical problem of food distribution.

16. **Fine medicines.** High-quality, affordable medicines, vaccines and pharmaceuticals may be made from algal coproducts or grown in algae bioengineered to produce advanced compounds such as antibiotics, vitamins, nutraceuticals and vaccines. These compounds are grown today in land plants and animals so algae offer significantly faster and lower cost production.

Designer algae grown locally in villages could save millions of lives by providing low cost vaccines or other medicines that need no packaging or distribution. Fine medicines, especially personalized drugs tailored to an individual, may offer more value than all other algal coproducts combined.

Local production will be enabled by training 10 million green masterminds distributed all over the globe who will be able to grow micro gardens of algae in a sustainable manner and at cost affordable to poor villagers, inner-city slums and back-yard gardeners.

Nature's first food production system on Earth, algaculture, offers extraordinary benefits. Solutions to commercial and micro-scalability combined with biotechnology and strain selection for digestible cell walls will ignite a green gold rush to produce high-value products especially food, fuels, fodder, fertilizers and medicines.

R3D; research, development, demonstration and diffusion requires substantial government investment to moderate risk during the startup phase. Production of new food and energy sources alone is an insufficient solution to the multiple constraints of climate change. Consumer behavior changes with movements such as the Green Fork Society will help people make eating and consumption choices that are ecologically friendly. Reducing demand for food products that require excessive land, water, fertilizers and fossil fuels will produce a very positive net yield to the environment.

Americans and world leaders are gifted in communication and celebration of success. When communities, states and countries develop goals and metrics for sustainable environmental resources such as water, carbon and pollution, achievement of those objectives can be monitored, recognized and rewarded. Similarly, innovators who enhance the capabilities to produce high-quality food and fuels at less cost will also be recognized and rewarded.

Solutions to mega-challenges such as world food, water and fuel security will be solved by a solution suite that includes a variety of renewable energy sources to supply the electrical power grid. Green Algae Strategy proposes that algae will be critical for providing food, valuable coproducts, pollution solutions and liquid transportation fuels. Successful achievement will require resources and a significant change in heart and habits.

Algae have promise to provide extraordinary solutions using three very different algal production scenarios:

1. **Large scale algal farms** will displace oil imports, enable energy independence and eventually end to the need to use fossil fuels with extensive help from other forms of carbon neutral fuels.
2. **Growing algae in large areas of the oceans** will sequester the $CO_2$ released from fossil fuels and halt global climate change.
3. **Micro scale algal gardens** will serve for ending hunger in America and the world and stop smoke death.

# Green Algae Strategy

The same R&D for algal production systems serves all objectives.

**Table 10.4 Strategies for Green Independence Objectives**

| Objectives | Strategy |
|---|---|
| End U.S. oil imports – 3600 days | • Build large algal farms to displace liquid transportation fuels.<br>• Engage other carbon neutral solar solutions to begin displacing fossil fuels for the electrical grid. |
| End U.S. use of fossil fuels – 20 years | • Continue the above actions and new carbon neutral energy comes on line to displace electrical grid.<br>• Algae continue to serve for needed liquid transportation fuels. |
| End fossil fuels globally – 40 years | • Continue the above actions and share knowledge and technologies. |
| Recapture carbon released in fossil fuels – 75 years | • Grow algae in massive ocean areas, especially dead zones and allow much of the carbon to sink through the water column. |
| End hunger in America – 10 years | • Design and build micro algal gardens to provide food, energy and other algal co-products locally.<br>• Train one million Green Masterminds in green solar gardens.<br>• Build micro gardens and train Green Masterminds in locations |

| | |
|---|---|
| | where people live in poverty and lack access to nutritious and delicious foods. |
| **End hunger globally – 30 years** | • Train ten million Green Masterminds globally in green solar production so they can grow what their families need locally.<br>• Create a Green Peace Corps to assist with training, building and gardening globally so families can produce for themselves locally. |
| **Stop smoke death (from fossil cooking fires) – 20 years** | • Ten million Green Masterminds address growing algae for cooking and heating fires first and for food and other co-products second. |

Algae saved our planet 3.5 billion years ago when it transformed the deadly $CO_2$ atmosphere to oxygen rich which supported life. Algae did what it does best, capture $CO_2$, put the carbon into green plant biomass and release pure oxygen to the atmosphere and oceans.

This humble plant stands ready to rescue the Earth again but his time it needs help from friends, Green Masterminds, because rescue must happen quickly or climate change will devastate human societies. Algae offer energy independence, a halt to global climate change as well as an end to hunger in America and the world.

These objectives are achievable but we will have to work together to help our fabulous friend; the oldest, tiniest yet fastest growing plant on Earth – algae.

## Acknowledgements

Thanks first to my best friend and life partner, Ann Ewen, who made this project possible. Ann supported both the "ah's" and the "aha!'s"

This product would not have been possible without the extraordinary research of Lester Brown, President of the Earth Policy Institute and Jeffery Sachs of the Earth Institute. Professors Qiang Hu, Milton Sommerfeld and Bruce Rittman supported endless questions on molecular biology and algae production. Environmental scientists Al Darzins, Eric Jarvis and Mike Siebert at NREL were very helpful with renewable energy sources. Oilgae author Narasimhan Santhanam served with great sources. Cindy Lucas and David Forsyth helped tremendously with editing. Thanks also to great advisors who elevated *Green Algae Strategy* from a solo to an orchestra.

| Science | Business – Econ. | Agribusiness |
|---|---|---|
| • Tom Moore | • Francine Hardaway | • Jon Ewen |
| • Jim Sears | • Mark Allen | • Gary Wood |
| • Brad Biddle | • Alan Resnik | • Doug Young |
| • Sam West | • Susan Schultz | • Jim Robertson |
| • Herb Roskind | • Gordon LeBlanc, Jr. | • Chris Kinsley |
| • Mark Ewen | • William Cockayne | • Barry Spiker |

Also helpful were the published works of Paul Ehrlich, Sandra Postel, David and Marcia Pimentel, Nobel Laureate Al Gore, Harvey Blatt, Fred Pearce, Michael Pollen and Linda Graham. High-content websites were a great support such as Algaebase, U.N., W.H.O., the National Resources Defense Council, Sierra Club, Green Peace, Audubon Society, Union of Concerned Scientists, Center for Energy and Climate Solutions, Clean Water Network and Public Citizen. Also useful were U.S. government sources including: DOE, EPA, USDA, NOAA and NREL.

# Mark Edwards

Mark has two goals he is pursuing through GreenIndependence.org:
1. End Hunger
2. Free us from oil imports

His vision: engage 10 million Green Masterminds globally who have the knowledge and capability for growing nutritious food and high-energy biofuel locally with green solar.

Algae's green promise, capturing carbon while producing food and biofuel offers green independence from oil imports for America. Green Independence engineers hope for millions of global citizens who lack access to affordable food, fuel and transportation.

Mark graduated from the U.S. Naval Academy in mechanical engineering, oceanography and meteorology where Jacques Cousteau motivated and mentored his interest in the oceans and global stewardship. He holds an MBA and PhD in marketing and consumer behavior and has taught food marketing, leadership, sustainability and entrepreneurship at Arizona State University for over 30 years.

Mark served as CEO of TEAMS Intl. for 24 years, the software and assessment firm he founded based on his research on advanced assessment technologies and talent and leadership assessment. He served as lead assessment and leadership consultant for over 400 firms globally. He was retained by many U.S. departments and the military, including DOE, DOD, Special Forces and the National Labs.

Mark served as a Director for a Fortune 50 foods company and has done extensive R&D on new foods, sources and consumer behavior. He has consulted for Monsanto, Pioneer Seeds, DuPont, Nabisco, Quaker Oats, General Mills, Borden and many other agribusiness companies. He has worked with senior executives at 15 large U.S. oil and gas firms as well as British Petroleum and Saudi Aramco.

## My Path to Green Independence

My green path began on a dairy and peach farm in central California where my dad taught me how to care for a wide variety of plants and animals and my mom gave me a passion for gardening. I worked alongside Braceros, Mexican field workers, growing and harvesting fruits, nuts, grapes and corn. Field work taught me the science and risks in agricultural production and the extraordinary pain of poverty.

Michael Harrington's *The Other America: Poverty in the United States* caught my imagination because I had experienced the poverty he wrote about. Our family was hard working farmer class but the Braceros and American farm workers lived in dismal poverty. Rachael Carlson's *Silent Spring* put me on an environmental track but the day after finishing her book, my job was to spray DDT on a fruit orchard.

My path elevated when Jacques Cousteau introduced a handful of Plebes at the U.S. Naval Academy to oceanography. We were encouraged to take new courses in oceanography and I took every course offered. The program was new and classes were small. Jacques Cousteau gave several lectures and conveyed the wonder of ocean discovery and global stewardship in his unique, charming manner.

After the Naval service, The Greyhound Corporation, a conglomerate with 152 subsidiary companies including Armour Foods and Dial, gave me a series of opportunities to use my oceanography knowledge to examine business proposals such as shrimp aquaculture, harvesting krill from oceans near Antarctica and other ocean foods. For a set of practical reasons, including competing with whales for food, my recommendation to the Corporation was to focus on food marketing and transportation and let specialist raise and harvest ocean foods.

My experiments with sustainability on an urban acre found the prospects about 80% possible but quite impractical due to the time requirements. Various forms of aquaculture, including algae for food and fertilizer turned out to be more mess than value. Without a scientific production template and more control over the algae species, home algaculture was impractical in the 1970s and 1980s.

For several decades I tracked algae as a food by visiting the library at Scripps yearly to read the few articles translated from Chinese, Japanese or Russian. This activity supported one lecture in my food marketing course on future foods where algae were one alternative.

Professor James Hershauer asked me to give a talk at church on the world food situation. Researching the lecture content aroused my concern about consuming food for biofuels. I read several hundred government reports and scientific articles and was deeply disturbed about the lack of science behind the U.S. energy policy.

Two excellent scientists who studied algae, Professors Milton Sommerfeld and Hu Qiang invited me to an ASU lecture by Al Darzins, Michael Siebert and Eric Jarvis from the National Renewable Energy Laboratory in Colorado. These scientists presented the case for algal biofuels but said that sadly, the U.S. government had cut off all funding for algae in 1993 and reassigned scientists to support the political favorite; corn ethanol.

My response was to research and write *Biowar I: Why Battles over Food and Fuel Lead to World Hunger* in order to move subsidies from corn to truly sustainable energy sources. *Biowar I* shares the unintended consequences of using an inefficient food source for fuel. Burning food for fuel depletes critical food supplies, drives up prices for all food inputs, creates food fights and will cause a food cascade that leads to the starvation of 30 million people. Continued ethanol production will leave America dry. Failing a policy change, a food cascade is 80% likely to occur in the next decade.

*Green Algae Strategy* conveys the value proposition for algae and is intended to benefit the millions of poor people who lack a voice in public policy and face starvation from lack of resources – cropland, freshwater, fertilizers and fossil fuels.

*Green Solar Gardens: Algae's Promise to End Hunger* describes the science and practice needed to distribute knowledge of algae production globally so people can grow sustainable and affordable food and energy locally for their needs. Green Solar Gardens addresses the root causes of poverty and hunger.

## Great Green Reading

*Food, energy and economics*

Thomas L. Friedman, *Hot, Flat, and Crowded: Why We Need a Green Revolution – and How It Can Renew America*, Farrar and Giroux, 2008.

Lester R. Brown, *Plan B 3.0: Mobilizing to Save Civilization*, Third Ed., W. W. Norton; 2008.

Jeffrey D. Sachs, *Common Wealth: Economics for a Crowded Planet,* Penguin Press HC, 2008.

Jeffrey D. Sachs, *The End of Poverty: Economic Possibilities for Our Time*, Penguin Press, 2005.

Fred Krupp and Miriam Horn, *Earth: The Sequel: The Race to Reinvent Energy and Stop Global Warming*, W. W. Norton, 2008.

Brangien Davis and K. Wroth, *Wake Up and Smell the Planet: Non-Preachy Grist Guide to Greening Your Day*, Mountaineers Books, 2007.

Daniel Esty and Andrew Winston, *Green to Gold: How Smart Companies Use Environmental Strategy*, Yale University Press, 2006.

*Water*

Elizabeth Kolbert, *Field Notes from a Catastrophe: Man, Nature, and Climate Change*, Bloomsbury, 2006.

Sandra Postel, *Pillar of Sand: Can the Irrigation Miracle Last?* W. W. Norton & Company, 1999.

Peter H. Gleick, *The World's Water 2006-2007*: *The Biennial Report on Freshwater Resources*, Island Press, 2006.

Fred Pearce, *When the Rivers Run Dry: Water –The Defining Crisis of the Twenty-first Century*, Beacon Press, 2007.

Vandana Shiva, *Water Wars: Privatization, Pollution, and Profit*, South End Press, 2002.

Robert Jerome Glennon, *Water Follies: Groundwater Pumping and The Fate Of America's Fresh Waters*, Island Press, 2004.

# Index

A2BE Carbon Capture, 168
100th Meridian, 49
Africa, 13, 19, 36, 62, 112, 195, 197
agar, 98, 142, 143
Agreenpromise.org, 183
agriculture, 12, 24, 74, 110, 126, 134, 154, 176, 177, 207
algaculture, 9 - 25, 51-75, 104-163, 172-220
algaculture limitations, 213
algaculture production 102
algal oil, 6, 7, 103, 124, 197
algal production, 18-25, 52, 70, 84, 114-134, 151-166, 176, 182-220
algal strains, 7, 14, 96, 106, 108, 111, 114, 156, 159, 162
Algenol Biofuels, 172
alginates, 117, 141
alginic acid, 79, 133, 140-144, 202
Mark Allen, 169
antibiotic compounds, 153
APS Redhawk, 102
aquaculture, 61, 89, 93, 134, 139, 152, 166, 192, 207, 211
Aquaflow Binomics, 167, 168
Aquatic Species Program, 82, 163
aquifer depletion, 45
aquifers, 10, 12, 42, 43, 44, 45, 46, 47, 50, 60, 62, 200
Arizona State University, 95, 108, 183, 201
atmosphere, 4, 8, 32, 43, 77-92
Australia, 13, 65, 81, 142, 174, 202
BASF, 162
John Beddington, 58
beta-carotene, 86, 142
Big Oil, 21, 208

biodiesel, 3, 6, 94-109, 150-183,
Biofuel Systems, 168
biomonitors, 159, 160
Bionavitas, 177, 181
biotechnology, 14, 70, 111, 117, 134-165, 172, 183, 199-220
Biowar I, i, iii, iv, 11, 33, 208, 226
Blue Marble Energy, 166
blue-green algae, 8, 78, 96-99, 136
Bodega Algae, 178
California, 13, 43, 46, 47, 50, 62, 111, 144, 157-164, 177- 200
Canada, 3, 150, 178, 202
carbon burial, 159
carbon neutral, 2, 3, 105
Carrageenans, 117, 139, 140
cell walls, 8, 41, 70, 79, 84-98, 108, 116, 124, 127-157, 198, 204,
Cellena, 178
cellulosic, 3, 6, 7, 8, 17, 18, 40, 107, 108, 197, 208
China, 3, 7-13, 36, 42-64, 142, 146, 180, 202, 210, 211
Chlorella, 90, 126, 127, 129
chlorophyll, 4, 77, 81, 85, 97, 148
climate change, 2, 13, 63
closed algaculture, 119
$CO_2$ levels, 32
collaboratory, iii, 183, 184
composition variation, 89
consumptive water, 43
coproducts, ii, 2, 15, 69, 133, 153, 157, 161, 188, 206, 209-220
coral reefs, 4, 79, 141, 182
cropland, 5-71, 109, 158, 189, 197, 215, 217, 218, 220, 221, 226

cyanobacteria, 8, 78, 80, 88, 99, 119, 139, 142, 153, 179, 183
Cyanotech, 180
Al Darzins, 40, 223, 224
DARPA, 200
dead zone, 34
demoisturing, 123
Desert Biofuels Initiative, 6
Diatomaceous Earth, 157
dinoflagellates, 79, 80, 112
Jacques Diouf, 29
disaster relief, 199, 220
DOE, 7, 21, 48, 82, 110, 182, 223
Dolphin Sea Vegetable Co, 180
dry cell weight, 98, 99
ecological suicide, 33
eflation, iii, 11, 30, 35, 36, 53
equity firms, 22
ethanol, 3, 6, 7, 11, 15- 54, 102, 153, 158, 163, 171-177, 184, 208, 209, 217, 218, 226
evaporation, 44, 51, 72, 74, 101, 109, 118, 140, 168
fat algae, 5
fatty acid, 98
fertilizer, 3, 5, 7, 11, 33, 41, 53-72, 81, 101-138, 156-158, 172, 177, 186-197, 204-206, 215, 218, 219
flagella, 79, 101
flocculants, 123
food assistance, 30
food cascade, iii, iv, 13, 42, 55, 226
food grains, 8, 10, 11, 17, 44, 52, 59, 128, 135, 152, 159, 217
food ingredients, 19, 105, 137-147
food riots, 11, 12, 33-37, 74, 191
food security, 27-29, 58, 65, 137,
forest death, 197
fossil fuels, 2, 3, 5-13, 34, 51-75, 150, 168, 175, 187-208, 218,

fossil water, 10, 12, 47-55, 69, 208
gag factor, 126, 130
glacier ice, 62
global Inst. of Sustainability, 183
Global warming, 2, 32, 59-63, 199
grain stores, 32
Green Algae Strategy 3, 10-25, 39, 73, 131, 161, 194, 207-226
green collar careers, 134
Green Fork Society, 191, 221
Green Revolution, 61- 62, 73, 128
green solar, iii, 2-8, 69, 118-129, 150, 185, 194, 209
Green Tag 10 Ecolo Footprint, 189
GreenFuel Technologies, 171
harvest, 103, 123, 212
Heifer Project, 72
herbicides, 3, 11, 33, 38, 39, 53, 54, 55, 60, 61, 63, 67, 69, 72, 73, 74, 107, 128, 159
James Hershauer, 194
HIV vaccine model, 201
Qiang Hu, 108, 223
hunger, 27, 28, 29, 35, 43, 66, 127, 147, 162, 185, 200, 207, 219
Hunger Project, 72
hydrogen, 3, 19, 77, 87, 92, 111, 135, 182, 218
immigrants, 37
industry language, 129
Ingrepo, 165
Inland algae, 145
Integrated system, 205
Inventure Chemical Tech, 174
Iowa State University, 40
irrigation, 10-12, 38-64, 74, 94, 103, 107, 119-128, 158, 176,
jatropha, 6, 15
jet fuel, 5, 7, 15, 18 , 69, 108, 135, 151-174, 187, 200, 218

Kelco, 166
kelp, 78, 95, 141-146, 166, 202-204
land plants, 17, 18, 78, 121, 139
LARB, 95, 108, 183
lichens, 78, 82, 86
Light Force, 149
lipids, 2-18, 53, 70, 83-108, 114-119, 131, 150-151, 183, 210
liquid transportation fuels, 6, 12, 18, 22, 52, 150, 209, 214, 218
LiveFuels, 15, 163
Mexico, 3, 19, 34-42, 47, 48, 92, 112, 154, 164, 165, 173, 180
microalgae, 78, 82, 83, 84, 124, 172, 176, 182, 227
mixing, 100, 114, 125, 212
Monsanto, 162, 180
Hiroshi Nakamura, 147
nanotechnology, iii, ii, 14, 137
NASA, 21, 127
National Corn Growers Assoc., 49
natural gas, 11, 35, 55, 62, 81, 159
Neptune Industries, 166
nitrogen, 7, 12, 34, 51-101, 112, 122-139, 145-159, 172, 183, 195, 202, 204, 206, 219
nitrogen fixing, 81
non-consumptive water, 45
Nori, 143, 144, 147, 202, 210
Nostoc commune, 179
NREL, 15, 21, 82, 110, 114, 223
nucleic acid, 136, 137, 198, 213
nutraceuticals, iii, 8, 20, 135, 151, 153, 179, 206, 214, 220
nutrient delivery, 114, 212
nutrient deprivation, 90
OAPEC, 10
ocean surface temp., 32, 60, 199
Ogallala aquifer, 10, 46, 47
Oglala Lakota tribe, 47

oil extraction, 124, 212
opportunistic invasion, 114
OriginOil, 164, 181
osmotic pressure, 79, 124
overdraft, 10
prsonalized drugs, 201, 202
pest vectors, 58, 59, 63
PetroSun, 164, 165, 181
pH, 72, 82, 95, 97, 99, 101, 116, 120, 127, 212
pharmaceuticals, 51, 131-134, 151-155, 188, 214, 220
photosynthesis, 5, 9, 69, 77, 80, 81, 84, 85, 86, 95, 97, 101, 118, 182
phycobilins, 142
phytoplankton, 78, 81, 82, 168
pigments, 5, 7, 8, 20, 85, 86, 100, 110, 12-142, 171, 179, 192
placebo effect, 148
plant biomass 5, 44, 63, 72, 81, 203
pollution, 3-19, 25-38, 54-75, 92, 109, 133-197, 208, 217, 221
pollution solutions, 74, 134, 158
poultry, iv, 39, 57, 145, 152, 190
poverty trap, 29
premier foods, 192
Prochlorococcus, 80
propagation, 17, 63, 90, 94, 155
R3D 23, 68, 113, **184**, 207, 208-211
really, really simple, 66, 68
red tides, 111
resource efficient, **67**
rising oceans, 61
robust, 66
rooftop algaculture, 187
Jeffry Sachs, 29, 68
save humanity, 126
Science Fiction, 128
Scripps Institution, 181
sea cows, 152

235

Seambiotic, 164
Jim Sears 169
seaweeds, 4, 88, 138-145, 180
sequester $CO_2$, 81
Josette Sheeran, 35
smoke death, iii, 196-200, 218
social networks, iii, 23
solar energy, 2, 5, 68, 126, 192
Solazyme, 172
Solena, 175
Solix Biofuels, 176
Milton Sommerfeld, 108, 226
South Africa, 3, 37, 42, 211
South America, 19, 20, 197
soybeans, 5, 8, 48, 64, 152, 197
*Soylent Green,* 129
space food, 138
Spirulina, 88, 90, 91, 93, 96, 97, 136, 142, 145, 146, 147, 148, 149, 179, 180, 186
storm surge, 199
strain selection, 14, 41, 106, 116, 117, 144, 220
subsidies, iii, iv, 21, 23, 24, 30, 34, 37, 39, 45, 49, 50, 53, 55, 110, 184, 207, 208
sulfur, 84, 90, 111, 146, 195, 219
sustainable communities, 191
sustainable food, ii, iii, 2, 23, 24, 61, 66, 68, 70, 71, 73, 75, 126, 194, 204, 208, 213
Sustainable Food Scorecard, 189
sustainable production, 12, 24, 25, 68, 186
Sustainable scorecard, 214

symbiosis, 4, 78, 86
terrestrial algae, 82, 92, 139
Texas, 39, 46, 47, 48, 78, 96, 117, 153, 164, 165, 175, 181, 182
texturized algal protein, 188, 193
texturized vegetable protein, 105, 131, 193
transpiration, 44
U.C. Berkeley, 78, 227
U.N. Report on Biofuels, 29
UNICEF, 72
unsaturated fatty acids, 91, 179
USDA, 21, 30, 34, 43, 48, 49, 223
value chain, 205, 206, 216
Vertigro Energy, 175
village scale algaculture, 186
Warren Belasco, 113, 128
wastewater, 17, 42-51, 101, 104, 121, 130, 158, 159, 217, 218
water tables, 12, 31, 47, 60, 62
wetlands, iii, vi, 5, 32, 33, 40, 44, 54, 60, 64, 65, 118
whales, 3, 111, 152
wine and beer, 192
Woods Hole, 81, 182, 227
World Bank, 35, 58
X Prize Foundation, 194
XL Renewables, 176, 177
Jean Ziegler, 35
Robert Zoellick, 35, 37, 58

[1] Interviews and surveys for the Green Algae Strategy project. 2007.
[2] Lewis, Leo. Japanese scientists create diesel-producing algae, *The Times*, June 14, 2008.
[3] FAO Report. Renewable biological systems for alternative sustainable energy production FAO Bulletin - 128, 1997.
http://www.fao.org/docrep/w7241e/w7241e0h.htm
[4] Ibid. 2.
[5] Huth, Hans. Biodiesel 101, 2008.
http://www.inkacola.com/greenbeat/soybenz/b101man/
[6] Eric Johnson and Brad Biddle founded the Desert Biofuels Initiative. http://desertbiofuels.blogspot.com/
[7] Vasudevan PT and Briggs M. Biodiesel production-current state of the art and challenges. *J. of Industrial Microbiology,* 2008, 35(5):421-30.
[8] Richmond, Amos. In A. Richmond, Handbook of Microalgal Culture, *Biotechnology and Applied Phycology*, Blackwell Science, Oxford, 2004, 4.
[9] Xia B. and I.Abbott. Edible seaweeds of China and their place in the Chinese diet, *Economic Botany*, 1987, 41: 341-353.
[10] Gressel J. Transgenics are imperative for biofuel crops. *Plant Science*, 2008, 174(3):246-63.
[11] Edwards, Mark. *Biowar I: Why Battles over Food and Fuel Lead to World Hunger*, Tempe: LuLu, 2008, 188.
[12] *Quality of Our Nation's Water*. Washington, DC: Environmental Protection Agency, 1994.
[13] Dias De Oliveira, Marcelo E., Burton E. Vaughan, and Edward J. Rykiel Jr. "Ethanol as Fuel: Energy, Carbon Dioxide Balances, and Ecological Footprint." *BioScience 55:7 (2005):* 593–602.
[14] Wee KM, Rogers TN, Hamm C. Engineering and medical applications of diatoms. *J Nanoscience and Nanotechnology,* 2005, 5(1):88-91.
[15] Kurki, Al, Amanda Hill and Mike Morris, Biodiesel: The Sustainability Dimensions, National Sustainable Agricultural Service, 2006, IP281.
[16] Dillehay TD, Ramírez C, Pino M, Collins MB, Rossen J, Pino-Navarro JD. Monte verde: Seaweed, food, medicine, and the peopling of South America. *Science,* New York, N Y. 2008, 320(5877):784-6.
[17] Edwards, Mark. Unpublished survey research.
[18] ArizonaCentral.com, Airlines push for homegrown jet fuel, August 15, 2008. www.azcentral.com/business/articles/2008/08/15/20080815biz

[19] Vesterby, Marlow, Kenneth S. Krupa. Estimating U.S. Cropland Area, *Amber Waves*, July, 2006, Special Issue.

[20] Chisti Y. Biodiesel from microalgae beats bioethanol. *Trends in Biotechnology*, 2008 Mar; 26(3):126-31.

[21] ENS. U.S. urged to merge land and oceans agencies into one, July 9, 2008. http://www.ens-newswire.com/ens/jul2008/2008-07-09-092.asp

[22] Jeffry Sachs, Jeffery. *Common Wealth: Economics for a Crowded Planet*, Penguin Press, 2008, 44.

[23] National Academy of Science. *Population Summit of the World's Scientific Academies*. Washington, DC: National Academy of Sciences Press, 1994.

[24] Black, Robert, Saul Morris, and Jennifer Bryce. "Where and Why Are 10 Million Children Dying Every Year?" *The Lancet* 361 (2003): 2226-34.

[25] Food and Agriculture Organization of the United Nations. *State of Food Insecurity in the World 2006*, Oct. 2006. ftp://ftp.fao.org/docrep/fao/009/a0750e/a0750e01.pdf.

[26] Ibid.

[27] Diouf, Jacques. *Revitalizing the Rural World: The Beginning of the End of Poverty*. Food and Agriculture Organization of the United Nations, Feb. 2006. http://www.fao.org/english/dg/oped/ICAARD.html.

[28] U.N. FAO, *Sustainable Bioenergy*, May, 2007, 3.

[29] Black, Robert, Saul Morris, and Jennifer Bryce. "Where and Why Are 10 Million Children Dying Every Year?" *The Lancet* 361 (2003): 2226-34.

[30] ERS, Food Assistance Landscape Report, 006http://www.ers.usda.gov/Publications/EIB6-4/

[31] Ibid.

[32] The years 2007 and 2008 are estimates.

[33] World Bank, *World Development Report 2008: Agriculture for Development*, October, 2007. http://publications.worldbank.org/ecommerce/catalog/

[34] Edwards, Mark. *Biowar I*.

[35] Experts quote 30% of U.S. corn crop in 2008. They forgot to account for the fact that refinery capacity will have doubled since 2006 and Mid-West floods lowered total corn production.

[36] Ibid., Chapter 7.

[37] *Linking Land Quality, Agricultural Productivity, and Food Security.* United States Department of Agriculture, Economic Research Service AER-823, 31. http://www.ers.usda.gov/publications/aer823/aer823e.pdf.

[38] Pegg, J.R. U.S. Corn Production Feeds Expanding Gulf Dead Zone, ENS, http://www.ens-newswire.com/ens/jun2008/2008-06-18-092.asp
[39] Sutter, John David. Five Water Bodies not Polluted, *News Oklahoma.com*, June 8, 2008, 1.
[40] Doyle, Alister, Ocean nitrogen only limited help for climate. Reuters, May 15, 2008. http://uk.reuters.com/article/oilRpt/idUKL153483820080515
[41] Edwards, Mark. Biowar I, 18.
[42] Rising Food Prices: Policy options and World Bank response, World Bank and FAO, April 9, 2008.
[43] Ten Kate, Daniel. Grain prices soar globally, *The Christian Science Monitor*. March 27, 2008, 1.
[44] UN: BiofuelProduction 'Criminal Path' to Global Food Crisis, April 28, 2008. http://www.ens-newswire.com/ens/apr2008/2008-04-28-03.asp
[45] Ibid.
[46] World Bank, 2008.
[47] Vital, John, *The Guardian*, April 05, 2008, 16. http://www.guardian.co.uk/environment/2008/apr/05/food.biofuels
[48] Editorial, World Food Crisis, *New Your Times*, April 10, 2008
[49] Ibid.
[50] Zoellick, Robert. High-Level Conference on World Food Security, Rome, World Bank, June 4, 2008. No: 2008/349/EXC.
[51] http://www.organicconsumers.org/articles/article_1371.cfm
[52] Food Planet, Food summit blames trade barriers, biofuels, June 4, 2008, www.planetark.com/dailynewsstory.cfm/newsid/48626/story.htm
[53] International Assessment of Agricultural Knowledge, Science and Technology for Development Report, April, 2008. http://www.agassessment.org/docs/SR_Exec_Sum_210408_Final.htm
[54] Tokgoz et. al., 2007, 39 and 44.
[55] Belmond and Nevada.
[56] Al Darzins, NERL, personnel correspondence, 2007.
[57] Chisti Y. Biodiesel from microalgae beats bioethanol. *Trends in Biotechnology*, 2008 Mar; 26(3):126-31.
[58] Diouf, Jacques. *Turning the Tide Against Water Scarcity*. Food and Agriculture Organization of the United Nations, Mar. 2007. http://www.fao.org/english/dg/oped/index.html.
[59] Gleick, P. *The World's Water 2001*. Washington, DC: Island Press, 2000: 52.
[60] Ibid., 52.

[61] Kijne, Jacob W. Unlocking the Water Potential of Agriculture. Rome: FAO, 2003: 26.
[62] California Farm Bureau. Water Education Foundation, Water Inputs in California Food Production, 1991.
[63] USDA, Water Use, http://www.ers.usda.gov/Briefing/WaterUse/
[64] Palmer, Tim. *Lifelines: The Case for River Conservation*, Washington: Island Press, 1996, 65.
[65] United States Department of Agriculture. *Economic Research Service, Irrigation and Water Use Briefing Room*, 2001. http://www.ers.usda.gov/Briefing/wateruse.
[66] Durwood, M., R. M. Dixon, and O. F. Dent. Consumptive use of water by major crops in Texas. Texas Board of Water Engineers, Bulletin 6019, 1960.
[67] USDA NASS, 2003, Farm and Ranch Irrigation Survey, Vol. 3. Special Studies Part 1, AC-02-SS-1, 176 http://www.nass.usda.gov/Census_of_Agriculture/2002/FRIS/fris03.pdf
[68] USDA, Farm and Ranch Irrigation Survey 2003.
[69] Pimentel, D., Berger, D, Filiberto, et al. Water Resources, Agriculture and the Environment, Environmental Biology, Report No. 04-1. New York State College of Agriculture and Life Sciences, 2004.
[70] Hutson, Susan S., Nancy L. Barber, et al. "Estimated Use of Water in the United States in 2000." *U.S. Geological Survey*, 2005. http://pubs.usgs.gov/circ/2004/circ1268/.
[71] Ashworth, William. *Ogallala Blue: Water and Life on the High Plains*, New York: William Norton, 2007, 13.
[72] Blatt, H. *America's Environmental Report Card: Are We Making the Grade?* Cambridge, Mass.: MIT Press, 2005: 71.
[73] McKinnon, Shaun. Worries Increase of San Pedro River's Health as It Runs Dry. *The Arizona Republic*, 29 June, 2007: B8.
[74] Hutson, Susan, et. al. "Estimated Use of Water in the U.S. in 2000, " USGA. http://pubs.usgs.gov/circ/2004/circ1268/htdocs/text-ir.html
[75] Gerry Sanders, Salt River Project, Personal interview, June, 2007. Three acre-feet is standard but some farmers get 5.5 acre-feet for their crops. The USDA Water Use Report, 2002, indicates and average of five acre-feet were delivered for the western states.
[76] National Corn Growers Association, Truths about Corn, Ethanol and Water Use, 2004. ncga.com/news/publications/PDF/GetTheFactsOnWaterUse.pdf
[77] USGS, Estimated Water Use in the U.S., 2000. http://pubs.usgs.gov/circ/2004/circ1268/htdocs/text-ir.html

[78] Personal correspondence, **2007**.
[79] Personal communication, **USGA**, 2007.
[80] Renewable Fuels Association, Ethanol Biorefinery Locations, 2008. http://www.ethanolrfa.org/industry/locations/
[81] Renewable Fuels Association, Ethanol Biorefinery Locations, 2008. http://www.ethanolrfa.org/industry/locations/
[82] Mark Edwards, Biowar I, 191.
[83] Today in Biofuels, *Biofuels Digest*, http://www.biofuelsdigest.com/blog2/2008/03/07/today-in-biofuels
[84] Zoellick, Robert. High-Level Conference on World Food Security, Rome, World Bank, June 4, 2008. No: 2008/349/EXC.
[85] Pearce, Fred. *When the Rivers Run Dry*. Boston: Beacon Press, 2006: 24.
[86] Ten, Kate, Daniel. Grain prices soar globally, *The Christian Science Monitor*. March 27, 2008, 1.
[87] AFP. Tibetan glacier melt leading to sandstorms, China Daily, 05-02-2006.
[88] Fred Pearce, Fred, 82.
[89] Mark Edwards, *Biowar I*, 76.
[90] Halweil, B. "Pesticide-Resistance Species Flourish." *Vital Signs*. Ed. L. Starke. New York: Norton, 1999: 124.
[91] "In Brief." *Environment*, Sept. 2001: 8.
[92] Bradsher, Keith and Andrew Martin. World's Poor Pay Price as Crop Research Is Cut, *New York Times*, May 17, 2008, B1.
[93] Ehrlich, P. R. and A. H. Ehrlich. *One with Nineveh: Politics, Consumption, and the Human Future*. Washington: Island Press, 2004: 71.
[94] DNR, "State of Iowa, Public Drinking Water Program 2006 Annual Compliance Report" (Des Moines, IA: June, 2007.
[95] Oster, Shai. "In China, New Risks Emerge At Giant Three Gorges Dam," *Wall Street Journal*, August 29, 2007, A1.
[96] Oster, A1.
[97] United Nations. *Report of the World Commission on Environment and Development*. General Assembly Resolution 42/187. 11 Dec. 1987.
[98] Sachs, Jeffery. *Common Wealth*, 44.
[99] Rogers, Everett. *Diffusion of Innovations*, 5th Ed., Free Press, 2003.
[100] This figure averages the estimates from 11 independent algae scientists who are producing algae.
[101] UTEX Algae Culture website. http://www.utex.org/
[102] Graham, L.E. The origin of the life cycle of land plants. *American Scientist*, 1985, 73; 78 – 96.

[103] Megasun.bch.umontreal.ca/protists/gallery.html algaebase.org/links/ utex.org; ccmp.bigelow.org; http://www.ccap.ac.uk; marine.csiro.au/microalgae; wdcm.nig.ac.jp/hpcc.html).

[104] Graham, Linda and Lee Wilcox. *Algae*. New Jersey: Prentice Hall, 2000: 8.

[105] Round, F. E. *The Ecology of Algae*. Cambridge University Press, 1981.

[106] Chisholm, Penny. The Invisible Forest: Microbes in the Sea, *MIT World*, 2008, http://mitworld.mit.edu/video/421/

[107] Hall, Jack. http://www.ecology.com/dr-jacks-natural-world/most-important-organism/index.html.

[108] IPCC. Bert Metz, Ogunlade Davidson, Heleen de Coninck, Manuela Loos and Leo Meyer (Eds.), Cambridge University Press, UK. 2005, 431.

[109] Hu, Qiang. "Environmental Effects on Cell Composition." *Handbook of Microalgal Culture: Biotechnology and Applied Phycology*. Ed. Amos Richmond. Oxford, England: Blackwell Science, Ltd., 2004: 83-94.

[110] Sheehan, John, Terri Dunahay, John Benemann and Paul Roessler, *NREL, A Look Back at the U.S. Department of Energy's Aquatic Species Program: Biodiesel from Algae, Close Out Report*, July 1998. http://www.nrel.gov/docs/legosti/fy98/24190.pdf.

[111] Sheehan, et. al., *NREL*: 4, 7.

[112] Ibid, 142.

[113] Ibid, 143.

[114] Masojidek, Jiri and Michael Koblizek. "Photosynthesis in Microalgae." *Handbook of Microalgal Culture: Biotechnology and Applied Phycology*. Ed. Amos Richmond. Oxford, England: Blackwell Science, Ltd., 2004: 20-39.

[115] Linda Graham, 12.

[116] Lewin, J. Diatom heterotrophy, General microbiology, 1953, 9: 305-313.

[117] van den Hoek, C., D.G. Mann, and H.M. Jahns. *Algae: An Introduction to Phycology*. Cambridge: Cambridge University Press, 1995.

[118] Graham and Wilcox, 7.

[119] My literature analysis shows a skew toward kill versus cultivation.

[120] Survey research, Morrison School, ASU, 2007, unpublished.

[121] Vermaas, William J. "Targeted Genetic Modification of Cyanobacteria: New Biotechnological Applications." *Handbook of Microalgal Culture: Biotechnology and Applied Phycology*. Ed. Amos Richmond. Oxford, England: Blackwell Science, Ltd., 2004: 380-91.

[122] Hu, Qiang. "Environmental Effects on Cell Composition." *Handbook of Microalgal Culture: Biotechnology and Applied Phycology*. Ed. Amos Richmond. Oxford, England: Blackwell Science, Ltd., 2004: 83-94.

[123] Huan M, Hamazaki K, Sun Y, Itomura M, Liu H, Kang W, Watanabe S, Terasawa K, Hamazaki T. Suicide attempt and n-3 fatty acid levels in red blood cells: a case control study in China.. *Biological psychiatr,* 2004, **56** (7): 490-6. PMID 1540784

[124] McNamara RK, Hahn CG, *et al.* Selective deficits in the omega-3 fatty acid docosahexaenoic acid in the postmortem orbitofrontal cortex of patients with major depressive disorder. *Biol. Psychiatry,* 2007, 62 (1): 17–24.

[125] Fan, Yang-Yi and Robert S. Chapkin. *Importance of Dietary γ-Linolenic Acid in Human Health and Nutrition. J. of Nutrition.* 1998, **128** (9): 1411-1414

[126] Johansen, J.R. Cryptogamic crusts of semiarid and arid lands of North America, *Journl of Phychology,* 1993, 29: 140-147.

[127] http://www.oilgae.com/algae/oil/biod/cult/cult.html

[128] Hu, Qiang. "Industrial Production of Microalgal Cell-Mass and Secondary Products–Major Industrial Species." *Handbook of Microalgal Culture: Biotechnology and Applied Phycology.* Ed. Amos Richmond. Oxford, England: Blackwell Science, Ltd., 2004: 264-73.

[129] Much of the content on algae attributes comes from the excellent *Algae* text by Linda Graham and Lee Wilcox, New Jersey: Prentice Hall, 2000.

[129] Round, F. E. *The Ecology of Algae.* Cambridge University Press, 1981.

[130] Graham, 24

[131] Guschina, I.A. and Harwood, J.L. Lipids and lipid metabolism in eukaryotic algae. Progress in Lipid Research, 2006, 45, 160–186.

[132] Sheehan, J., Dunahay, T., Benemann, J. and Roessler, P.G. US Department of Energy's Office of Fuels Development, July 1998. A Look Back at the US Department of Energy's Aquatic Species Program – Biodiesel from Algae, Close Out Report, 1998, TP-580-24190.

[133] Qiang Hu, Milton Sommerfeld, Eric Jarvis, Maria Ghirardi, Matthew Posewitz, et. al. Microalgal triacylglycerols as feedstocks for biofuel production: perspectives and advances , The Plant Journal, 2008, 54 (4), 621–639 doi:10.1111/j.1365-313X.2008.03492.

[134] Hu, Q., Zhang, C.W. and Sommerfeld, M. Biodiesel from Algae: Lessons Learned Over the Past 60 Years and Future Perspectives. Juneau, Alaska, Phycological Society of America, July 7–12, 2006, 40–41.

[135] Durrett, T., et. al. Plant triacylglycerols as feedstocks for the production of biofuels. Plant Journal, 2008, 54, 593–607.

[136] Renaud, S.M., Thinh, L. Effect of temperature on growth, chemical composition and fatty acid composition of tropical Australian microalgae grown in batch cultures. Aquaculture, 2002, 211, 195–214.

[137] Alonso, D.L., Belarbi, E.H., Fernandez-Sevilla, J.M., Rodriguez-Ruiz, J. and Grima, E.M. Acyl lipid composition variation related to culture age and nitrogen concentration in continuous culture of the microalga *Phaeodactylum tricornutum*. Phytochemistry, 2000, 54, 461–471.

[138] Mansour, M.P., Volkman. The effect of growth phase on the lipid class, fatty acid and sterol composition in the marine dinoflagellate, *Gymnodinium* sp. in batch culture. Phytochemistry, 2003, 63, 145–153.

[139] Tomasellin, Luisa. "The Microalgal Cell." *Handbook of Microalgal Culture: Biotechnology and Applied Phycology*. Ed. Amos Richmond. Oxford, England: Blackwell Science, Ltd., 2004: 3-19.

[140] Zmora, Oded and Amos Richmond. "Microalgae for Aquaculture." *Handbook of Microalgal Culture: Biotechnology and Applied Phycology*. Ed. Amos Richmond. Oxford, England: Blackwell Science, Ltd., 2004: 364-79.

[141] Chen C, Zhang X, Zhu L, Liu J, He W, Han H. Disinfection by-products and their precursors in a water treatment plant in north china, *The Science of the Total Environment*, 2008, 397(1-3):140-7.

[142] Hu, Industrial Production, 264-73.

[143] Green Fuels Technologies. http://www.greenfuelonline.com/.

[144] Masojidek, Jiri and Michael Koblizek. "Photosynthesis in Microalgae." *Handbook of Microalgal Culture: Biotechnology and Applied Phycology*. Ed. Amos Richmond. Oxford, England: Blackwell Science, Ltd., 2004: 20-39.

[145] Hu, Environmental Effects, 83-94

[146] Hu, Environmental Effects, 83-94.

[147] http://www.aps.com/general_info/newsrelease/newsreleases/

[148] http://www.greenfuelonline.com/technology.htm.

[149] Lantz, M. *et al.*, The prospects for an expansion of biogas systems in Sweden, *Energy Policy* 35, 2007, 1830–1843.

[150] Sánchez Mirón, A. *et al.*, Biochemical characterization of microalgal biomass, *Enzyme Microbiological Technology*. 31, 2002, 1015–1023.

[151] Hu, Qiang. Personal communication, June 2007.

[152] NREL, "A Look Back at DOE's Aquatic Species Program: Biodiesel from Algae," 1998, iii. http://www.nrel.gov/docs/legosti/fy98/24190.pdf

[153] Richmond, ed. Handbook of Microalgal Culture.

[154] Schmidt W, Bornmann K, Imhof L, Mankiewicz.. Assessing drinking water treatment systems for safety against cyanotoxin breakthrough using maximum tolerable values. *Environmental Toxicology*, 2008, 23(3):337-45.

[155] UTEX Algae Collection, http://www.utex.org/

[156] Ugwu CU, Aoyagi H, Uchiyama H. Photobioreactors for mass cultivation of algae. *Bioresource Technology*, 2008, 99(10):4021-8.
[157] Florida Department of Environmental Protection, www.bioreactor.org/
[158] Meier, R. L. industrialization of photosynthesis and its social effects, Chemical and Engineering News, October 24, 1949, 3112 – 16.
[159] Velie Food pumpkin pipelines, 9 – 14, Referenced in Warren Belasco. *Meals to come,* Berkeley: University of California press, 2006, 204.
[160] Brockman, M. C., A. Henic. Biosorption of heavy metals by Fucus spiralisk, Kurtz and Tischer. Closed cycle biological systems for space feeding. *Food Technology*, 1958, 12: 449. Coleman, Herbert J., U.S., Russian scientists view algae as principal space food, *Aviation Week and Space Technology*, August 14, 1967, 88 – 89.
[161] Belsco, Warren. *Meals to Come: A History of the Future of Food,* Berkeley: University of California press, 2006, 106.
[162] Harrison, Harry. Make Room! Make Room!, New York: Berkley Medallion, 1966, 3, 131, 173.
[163] Flavr Savr is the registered trademark of the Calgene Corporation and Alnuts is the registered trademark of the Alnuts Foundation.
[164] Xia and Abbott, Algae as Food, in Lembi, Carole and Waaland, Robert. *Algae in human affairs,* Cambridge University Press, 1988.
[165] Lembi, Carole and Waaland, Robert. *Algae in human affairs,* Cambridge University Press, 1988.
[166] Richmond, A. Spirulina in *Micro algal biotechnology*, edited by M.A. Borowizka and L.J. Borowizka, Cambridge University, UK. 1988, 85 – 121.
[167] Kofranyi, E. The nutritional value of the green alga scenedesmus acutus for humans, Ergebn limnol., 1978, 11: 150.
[168] Chamorro, G. Etude toxicologique de l'algue *Spirulina* plante pilote productrice de proteins, UF/MEX/78/048, UNIDO/10.387 (1980).
[169] Indergaard, M. and Minsaas, J. 1991. Animal and human nutrition. *in* Guiry, M.D. and Blunden, G. 1991. *Seaweed Resources in Europe: Uses and Potential.* John Wiley & Sons.
[170] Kamarei, A.R. et. al. Potential for utilization of alga biomass for components of the diet in CELSS, Society of automotive engineers, technical paper series, number 851388. 15th Intersociety conference on environmental systems, San Francisco, 1985, July, 15 – 17.
[171] Mumford, Thomas and Akio Miura,. Porphyoa as food: cultivation of economics, in *Algae in Human Affairs*, Carol Lembi and Robert Waaland, Eds, Cambridge: Cambridge University press, 1988, 87 – 117.

[172] Jassby., Alan. Spiralulina: a model for microalgae as human food, in *Algae in Human Affairs*, Carol Lembi and Robert Waaland, Eds, Cambridge: Cambridge University press, 1988, 149 – 173.

[173] Wernberg T and Goldberg N. Short-term temporal dynamics of algal species in a subtidal kelp bed in relation to changes in environmental conditions. *Estuary Coast Shelf Science*, 2008, 76(2):265-72.

[174] Becker EW. 2007. Micro-algae as a source of protein. *Biotechnology Advances*. 25; 2, 207-10.

[175] Kram, Jerry. Seeking Cyanobacterial Cellulose, *Biomass Magazine*, June, 2008, www.biomassmagazine.com/article.jsp?article_id=1679

[176] Gerwick,W.H. et. al. Screening cultured marine algae for anticancer-type activity. Journal of applied phycology, 1994, 6: 143-149.

[177] Pushpamali, Wickramaarachchilage Anoja, et. al.. Isolation and purification of an anticoagulant from fermented red seaweed Lomentaria catenata, *Carbohydrate Polymers*, 7; 2, July 2008, 274-279.

[178] UPI.com. Oil drilling may help biomedical research, July 3, 2008. http://www.upi.com/Science_News/2008/07/03/

[179] Bock R: Plastid biotechnology: prospects for herbicide and insect resistance, metabolic engineering and molecular farming. *Current Opinions in Biotechnology*, 2007, 18.

[180] Dellapenna D, Pogson B: Vitamin synthesis in plants: tocopherols and carotenoids. *Annual Review of Plant Biology*, 2006.

[181] Mayfield SP, et.al. 2007. Chlamydomonas reinhardtii chloroplasts as protein factories. Current Opinioins in Biotechnology, 18;2, 126-33.

[182] González, E. Romera, F., A. Ballester, M.L. Blázquez, J.A. Muñoz. Biosorption of heavy metals by Fucus spiralis, *Bioresource Technology*, 99; 11, July 2008, 4684-4693.

[183] Linda Graham 28.

[184] Lewis, M.A. Are laboratory derived toxicity data for freshwater algae worth the effort? *Env. toxicology and chemistry*, 1990, 9: 1279 – 1284.

[185] See www.BiofuelsDigest.com.

[186] Weiss, Rick. Firms Seek Patents on 'Climate Ready' Altered Crops, Tuesday, May 13, 2008; Page A04ps, Washington Post, May 13, 2008; A4.

[187] Marrero, Carmelo Ruiz. Biotech Bets on Agrofuels, Center for International Policy (CIP), April 24, 2008. http://americas.irc-online.org/am/5179

[188] Solazyme web site, http://www.solazyme.com/news080415.shtml
[189] Patent application, www.wipo.int/pctdb/en/wo.jsp?wo=2007070452&IA=WO2007070452
[190] Colonies of nostoc commune: methods for cultivating edible nostoc commune and edible nostoc commune formulations and their use for promoting health, http://www.freepatentsonline.com/70160704.html
[191] Fox, R. *Algoculture*, Doctoral thesis. Louis Pasteur University, Strasburg, 1984, 336.
[192] Chaudhari, P.K. et. al. Growth potential for Spirulina, blue-green alga, in night soil digester effluent and saline water. I*ndian Journal of environmental health*, 1983, 25, 75 – 81.
[193] DARPA, Biofuels. http://www.darpa.mil/sto/solicitations/BioFuels/
[194] Steve LeVine, *Business Week*, April 28, 2008. Businessweek.com/bwdaily/dnflash/content/apr2008/db20080428, 5
[195] WHO, *Global Burden of Disease*, 2000.
[196] Ibid., 2000. www.who.int/indoorair/health_impacts/burden_global/en/index.html
[197] WHO, http://www.who.int/indoorair/en/
[198] Pictures from the WHO web site.
[199] http://www.spirulinasource.com/algaemyanmar.html
[200] Balzarini J. 2006. Inhibition of HIV entry by carbohydrate-binding proteins. *Antiviral Research,* 2-3, 237-47.
[201] International Assessment of Agricultural Report, April, 2008.
[202] Mark Edwards, *Biowar I*. 212.
[203] http://genome.jgi-psf.org/mic_cur1.html); http://genome.jgi-psf.org/euk_cur1.html.
[204] Charrier B, Coelho SM, Le Bail A, et al. Development and physiology of the brown alga ectocarpus siliculosus: Two centuries of research. *New Phytologist*, 2008, 177(2):319-32.
[205] Ibid. 322.
[206] Ibid. 328.

Green Algae Strategy

Made in the USA